The Talking Wire

This edition published 2022
by Living Book Press

ISBN: 978-1-922919-16-8 (hardcover)
        978-1-922919-15-1 (softcover)
        978-1-922919-16-8 (ebook)

All rights reserved. No part of this publication may be reproduced, stored in a retrieval system, or transmitted in any other form or means – electronic, mechanical, photocopying, recording or otherwise, without the prior permission of the copyright owner and the publisher or as provided by Australian law.

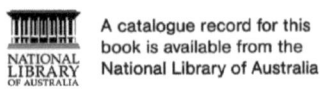

A catalogue record for this book is available from the National Library of Australia

THE TALKING WIRE
The story of Alexander Graham Bell

BY
O. J. STEVENSON

ILLUSTRATIONS BY
LAWRENCE DRESSNER

Living Book Press

Contents

Part 1.

i.	The Paddle Wheel	3
ii.	"A Very Intelligent Dog"	10
iii.	"Graham" Bell	15
iv.	Grandpapa Bell	18
v.	The Great Adventure	21
vi.	"The Little Man"	28
vii.	"Visible Speech"	36

Part 2.

viii.	A Career	43
ix.	The Enchanted Lyre	49
x.	The Alarm	53
xi.	Tutela Heights	59
xii.	The War Dance	65

Part 3.

xiii.	The Plunge	71
xiv.	The Magic Glove	78
xv.	"A Real Ear?"	83
xvi.	The Dreaming Place	87
xvii.	"A Little Mad"	90
xviii.	"Get It!"	96
xix.	Stress and Strain	99
xx.	The Twang of the Wire	102

| xxi. | A Crisis | 108 |
| xxii. | March 10, 1876 | 114 |

Part 4.

xxiii.	The Emperor	121
xxiv.	Stovepipe Wire	126
xxv.	The Shuttle	132
xxvi.	In England	135
xxvii.	The Telephone, Past and Present	141
xxviii.	At Home and Abroad	144
xxix.	Beinn Bhreagh—Beautiful Mountain	147
xxx.	"No End to Striving"	150
xxxi.	The Reminder	155
xxxii.	At Close of Day	162

Part I.

BOYHOOD DAYS.

CHAPTER ONE

The Paddle Wheel

Everyone in Edinburgh knew where Herdman's big flour mills were, out at the edge of the city, ten or fifteen minutes' walk from Princes Street Gardens. On a September afternoon, Mr. Herdman, the miller, was standing before a tall office desk, looking over his accounts. Two boys, Ben Herdman and Aleck Bell, were "playing around" somewhere, and, as like as not, they were into some mischief. He could hear them shouting to each other over the noise of the mill. But all of a sudden they became very quiet—usually a sign that some mischief was brewing—and then a different sound caught his ear. It seemed as if the mill-wheels were going to stop. Something had gone wrong! He rushed out of his office, banging the door behind him in his haste. Those boys again! Sure enough, the two mischievous young imps were trying to close the gates and shut off the water. They wanted to see what would happen—and they did! They were always up to something! But this was the worst ever!

When they caught sight of Mr. Herdman, they scuttled away and tried to hide; but when he had opened the gates once more, he routed them out. He was angry, but he didn't quite lose his temper.

"You young rascals!" he shouted. "Come out of that! Haven't I warned you a hundred times? Can't you read? See that sign,

'Keep out'—that means *you*! Come over here to my office. I want to talk to you. You, Ben, you ought to know better. What do you mean?"

"I was just showing Aleck," pleaded Ben.

"Aleck!" shouted Mr. Herdman. "His father'll attend to him. Serve him right, too."

Aleck's heart sank. Mr. Herdman *might* tell his father. He hadn't thought of that. If he did, he knew what would happen! He wouldn't be allowed to come to the mill again.

"I'm sorry, sir," he said. "It wasn't Ben's fault. I dared him."

Mr. Herdman began to relent. "You know," he said, "I can't be watching you boys all the time. You're giving me no end of trouble. This isn't the first time, but it must be the last. Why can't you do something useful instead of getting into mischief all the time?"

"What is there we *could* do?" ventured Aleck.

"Do?" said Mr. Herdman. "There are a hundred things for you to do—and a hundred things not to do. Look at this!" And, suiting the action to the word, he thrust his hand into a sack of grain that was standing near by, and brought out a handful of wheat and spread it out before them.

"See that wheat?" he said. "See anything wrong with it?"

Aleck looked at it. "You couldn't grind that, could you, sir? They need the—the—"

"Husks," prompted Mr. Herdman.

"They need the husks taken off, don't they?"

"Right," said the miller. "But how? If you can find some way to do that, you'll be a help instead of a hindrance."

"Why don't the husks come off when they're flailed?" asked Ben.

"Sometimes the wheat is not dried out enough before they're flailed," replied Mr. Herdman. "Some kinds of wheat are like that. You can't shake the husks off or rub them off or blow them off."

"If you'll let me take some home," said Aleck, "I'd like to try—two or three handfuls'll do."

"Well," replied Mr. Herdman, "it won't do you any harm to try! Good luck! And don't you meddle with that water again! Or—" The rest of the sentence was lost behind the closed door.

Ben and Aleck were the same age, "eleven, going on twelve." Ben stuttered, and his father had taken him to Professor Bell, the well-known teacher of speech, to see if he could do anything for him, and now he was going to Professor Bell's house once each week for a lesson. That was how Ben and Aleck came to know each other. The Bells lived on South Charlotte Street, not very far from the centre of the city, in one of those large apartment houses in which there were some times a half-dozen families. It was a dull place, and very noisy too; and the family, Aleck and his two brothers, Melville and Edward, didn't very often get out to the open country; and for Aleck an afternoon at Herdman's mill was something to look forward to and dream about. Anything to get away from the city streets and the smoke-stained, grimy stone houses and cobbled pavements!

On this particular afternoon when Aleck reached home he found his brother Melville waiting for him.

"Time you were home," said Melville. "Mama's been looking for you. You haven't done your practice and to-morrow's your lesson. Signor Bertini—"

"Practice!" interrupted Aleck. "I've done enough practice this week. I've got something else to do right now. See here!" He opened the grain sack cautiously, as if a mouse might jump out, and held it up for Melville to see. Melville peered in curiously, and took out a handful.

"Pooh!" he said scornfully. "Just wheat, isn't it? What'll you do with it? Plant it?"

"No," said Aleck. "I've nowhere to plant it, you know that, but I have to find some way to take those husks off."

"That's easy," said Melville. "Rub it."

"No," said Aleck. "That won't do. You just try it. It's too slow. Mr. Herdman has piles and piles of it. You'd never get it done."

Just then a voice came from an open window. "Aleck, you must get at your piano. You won't be ready for your lesson."

Aleck didn't answer. It was no use. His mother was too deaf to hear. He didn't mind practising. He had a good ear and could play almost anything if he heard it once. He had a good music teacher too—the best in Edinburgh—Signor Auguste Benoit Bertini. Aleck's father was careful in money matters, but he spared nothing for the education of his three boys, and Aleck was an apt pupil. He was music to his finger tips; and he had a vague notion that when he grew up he would be a teacher like Signor Bertini, and play before great crowds of people in London. Signor Bertini would be proud of him.

Aleck went to the piano. He played "The Girl I Left Behind Me"—a catchy sort of thing that everyone was singing—then he ran over his lesson once; but he spent most of his time making up pieces—"improvising" Signor Bertini called it—and he forgot all about Mr. Herdman's grain. The music put everything else out of his head. It always did! But that evening he took the little sack of wheat and spread it out on the table. He ran his hands through it. He liked to feel it slipping through his fingers. He rubbed out a handful, but it was slow work and the wheat wasn't very clean after all. Then a new idea came to him. "Why can't I brush the husks off," he said to himself, "the way I brush the mud off my boots? All I need is a good stiff brush." He got an old nail brush from a bureau drawer, and tried it. The husks came off quite easily and he blew the chaff away. He worked till he had two or three cupfuls—but at that rate of going it would take all day to do a bushel —and what would Mr. Herdman say? He wished he had left those gates alone!

Two or three days later he hurried down to the mill again, and showed Ben the wheat he had husked, and he brushed some more of it to let him see how it was done.

"Neat, isn't it?" said Ben. "Let *me* try it."

They took turns with the brush, and soon had a little pile for Mr. Herdman to see. But it wouldn't do. It was too slow and it was very tiresome work.

"I want to find some way to clean up a hundred bushels at a time," said Mr. Herdman. "It would take all summer to do it with that little brush."

"Why don't you get a bigger one?" said Ben.

Aleck didn't know where to find one, and he couldn't think of anything else. But suddenly a bright idea came to him. Outside in an out-of-the-way corner of the yard there was an old vat, lined with wire or some rough material. Inside it there was a paddle-wheel, one that went round and round when you turned the crank. No one knew what it had been used for, but when Aleck happened to see it, he gave a whoop of delight.

"Look!" he exclaimed. "Here's the very thing! Put your wheat into it—a whole sackful—now let's paddle her!"

No sooner said than done! Round went the paddle. The wheat was splashed up against the wire, and the husks came off like magic.

"Hurrah," cried Aleck. "What'll your father say now?"

"He'll not say much," said Ben. "He never does."

Mr. Herdman didn't say much at first; but when he saw the paddles working, he put his hand on Aleck's shoulder and said, "Well done, my lad! Well done!" and Aleck was very proud. Mr.

"Well done, my lad! Well done!"

Herdman had a bigger vat made, and a paddle-wheel that went by water power; and for many a long day the wheat at Herdman's mills was cleaned by the paddle-wheel method.

"Clever boy of yours—that boy Aleck," said the miller when he met Professor Bell a fortnight later. "You know, he's found a way to take the husks off our wheat when the flail won't do it. Came to him quick as a flash. He'll be an inventor someday, if I know anything. A little mischievous—but not enough to amount to anything. How does he get along at school?"

"He's idle," said Aleck's father, "dreaming half the time! He likes arithmetic, but he hates Latin, and won't study Greek, like Sir Walter Scott when he was a boy. 'The Greek blockhead,' they called him."

"I don't blame him," said Mr. Herdman. "What's Greek good for, anyway? Send him down to me and I'll make a miller of him."

But Aleck's father had other plans for him, that were quite different!

CHAPTER TWO

"A Very Intelligent Dog"

THE BELL FAMILY were teachers of speech, Grandpapa Bell, who lived in London, and his two sons, Uncle David, who taught in the University at Dublin, and Aleck's father, Professor Alexander Melville Bell, who lived in Edinburgh. In Professor Bell's family there were three children, Melville James (Melly), Aleck, and Edward Charles (Ted), three bright boys. Their mother was a woman with unusual gifts. She was a skilled painter of miniatures and gave lessons in painting. In spite of the handicap of deafness she excelled as a pianist and after the death of Signor Bertini, Aleck's music teacher, she taught the boy; and in addition to these accomplishments she was always actively interested in everything that her husband and her sons did.

Besides the boys and their parents there was another member of the household who must not be forgotten, the little dog who bore the odd name of "Mr. Perd." He was a little black Skye terrier with a spot of white on his breast, a long body and short legs and long hair that straggled down over his eyes. Perd belonged to the family, but he was Aleck's special care; and the two of them went everywhere together. It was Aleck who fed him and combed his long hair and taught him all he knew. He soon learned all the common tricks, to beg and shake hands, and to lie down and be a "dead dog."

South Charlotte Street, where the Bells lived, was not the best place for either a boy or a dog. The house was gloomy and the street was noisy; and they spent as much time as they could at a cottage—Milton Cottage by name—in the open country near Edinburgh. There in the long summer afternoons Aleck undertook to teach Perd something new.

Since their father was an expert in curing persons of defective speech, the boys heard a good deal about the use of the tongue and lips and vocal cords in speaking. This set Aleck thinking, and one day he said to his father, "Papa, why can't Perd speak?"

"The real reason why he doesn't," said Professor Bell, "is that he hasn't anything to say. He speaks with his ears and his eyes and his tail, and he knows how to beg; and when there is anything exciting going on, he barks or growls or whines, and sometimes he rushes around as if he had gone completely out of his mind. He understands what you mean when you say, 'Down' to him, but he can't say 'Down' to you. He doesn't need to!"

"Well," said Aleck, "I wish he'd say something. There's Mrs. Jones' parrot. He can talk, and he's not half so clever as our dog. Couldn't we teach Perd one or two words?"

"No, I don't think so," said Professor Bell, "but you may try if you like. Teach him to stand on his hind legs and growl. That'll do to begin with. When he growls ever so little, give him a biscuit or something else that is good to eat. If you do that often enough, by and by he'll do nothing but growl. If you feed him and praise him when he does well, you won't have much trouble. But you have to be patient with a dog until he finds out what you really want. But I've forgotten the most important thing after all. What does he say when he growls? What does it sound like? Is it 'Oh' or 'Ah' or 'Ee,' or is it something else?"

"It sounds more like 'Ah' than anything else," said Aleck. "'Ah! Ah! Ah!' He just keeps on saying 'Ah'."

"Now then," said Professor Bell, "that's enough for one lesson. But make him keep on growling, and don't feed him unless he does."

The next time that "Mr. Perd" had a lesson, Professor Bell was the schoolmaster. As soon as Perd began to growl, Mr. Bell took hold of the dog's muzzle, and opened and closed the jaws three or four times, one after the other.

"Listen," he said to Aleck. "What's he saying now?"

"He's saying, 'Ma-ma, Ma-ma'," said Aleck.

"Now then," said Professor Bell. "You try it. Don't be afraid. He won't bite. He wants that cake—that's all! Now make him do it again. A dog doesn't learn very quickly. He has to get the idea, and he has to say it over and over and over again, the way you do with your Latin."

"Latin!" laughed Melly, who happened to come up just then. "How do you learn your Latin, Aleck? *Amo, amas, amat.* You ought to get an extra piece of cake for that!"

Aleck said nothing, but he gave his brother a withering glance.

"One thing at a time," said Professor Bell. "Don't give up too easily. Give him lots of practice. He must get used to it. And remember, no growl; no cake!"

The next lesson was harder, and Perd didn't like it, and growled all the louder.

"I want to shut the air off," said Professor Bell, "just the way you shut the water off a tap. I'll show you how to do it. Press up on his throat between the lower jaw bones, while he is growling, and make him say, 'Ga! Ga! Ga!' Here Melly, your turn now, over and over and over again, the way you learn your Latin. That'll

need some practice. When he has learned that, come back to me again. There is just one lesson more."

This last lesson wasn't so easy, either; but Aleck soon learned what to do. The dog's lips had to be held together, narrowed and rounded, so that he made the sound "Ow" (or "Ah-oo"). That took practice and extra biscuits, but by and by the little dog growled through the narrow opening and said "Ah-oo" in good style.

And now that "Mr. Perd" had learned a new kind of alphabet, the boys tried to put these different sounds together to see if they would make sense; and after trying this arrangement and that, without success, they finally made up a sort of sentence that did say something. It wasn't "dog-latin" or "doggerel," and it sounded like baby-talk; but with a good imagination you could make out something like this:

"Ow-ah-ooaga-mama?" or in other words, "How are you, Grandmama?"

It was, of course, all very childish, but it answered Aleck's question whether a dog could learn to speak.

"That's all," said Aleck's father. "It isn't even dog language, but there isn't anything better we can do."

But Aleck wasn't quite satisfied and kept on drilling Perd on these sounds until he could say them almost as fast as Aleck could move his fingers. Perd, of course, didn't know anything about "Ga-mama," and he thought all the time that he was saying, "Come on with your cake." Poor Perd made another sad mistake. He seemed to think that he could ask about Grandmama's health without the help of Aleck; but no matter how hard he tried, he got the sounds all mixed up. He couldn't say "Ow-ah-oo" by himself any more than Aleck could speak Greek.

In the meantime Perd was becoming quite famous. You can't have a dog that talks, without the neighbours knowing all about it. Visitors came from far and near to see him, and they all agreed that he was a very intelligent dog.

CHAPTER THREE

"Graham" Bell

ONE AFTERNOON WHEN Aleck returned from school, he found his father closeted with a visitor.

"Who is he?" asked Aleck of his brother Melly. "What's wrong with him? Does he stutter?"

"No," said Melly, "there's nothing wrong with him. He's a friend of Papa's, that's all. Papa went to school with him. He lives in London now, but he owns plantations in the West Indies, Cuba, I think. He has just called to see Papa on his way back home."

"What's his name?" asked Aleck.

"I'm not sure," said Melly. "Graham, I think. Father calls him Aleck."

Aleck made a wry face. "Another Alexander? I wish that Papa and Mama had called me something else."

"Well," said Melly, "I don't see what you can do about it. You're named after Grandpapa, and you can't change it."

"I don't mind the name," said Aleck, "but I wish there weren't three of us, Alexander I, Alexander II, and Alexander III. I don't want to be mistaken for Grandpapa. If I had two names, like you, I wouldn't mind. If you don't like 'Melville' or 'Melly' you can change it to 'James.' But I'm just plain 'Alexander.'"

Mr. Graham, Professor Bell's friend, did not stay long; but Aleck liked him, and he liked the name "Graham." He would soon be eleven (Grandpapa's birthday and his fell on the same day). Why not take the new name "Graham" then? The more he thought of it, the more he liked it; and he told his brother Melly about it.

"Aleck says he's going to change his name," shouted Melly to his mother.

"Nonsense, child," she exclaimed, a little startled at the thought. "Another of his notions! What's the matter with the boy? Change his name? You're surely not in earnest?"

She was a little dismayed. She didn't really object to the name "Graham"; but Aleck was showing his independence, and that was a disturbing thought.

"It's a good enough name," said Aleck's father. "Alexander Graham is an old friend of mine and he would be greatly pleased if you took his name; but "Alexander" is a sort of family name. There have always been Alexanders in the Bell family, and what would Grandpapa say? He would be very much hurt. No, you mustn't do that, not while your grandfather is alive anyway."

They talked it over, father and mother and sons, the next day and the next, whenever they were together. "You can't blame the boy," said his mother, "for wanting to have a name of his own. It's the three 'Alexanders' that he doesn't like. Let him be 'Graham' to other folks if he wants. He'll always be 'Aleck' to us."

"Well, do as you like," said Mr. Bell turning to Aleck, "but don't talk about it before Grandpapa. Remember, you're his namesake."

In the meantime his two brothers gave him no peace. When they wanted to tease him they called him "Alexander the Great", and that made him furious. They did not know that someday he would be "Alexander the Great". You don't have to have an army, to conquer the world.

Anyway, he wasn't Alexander Bell any more, except to his

grandfather. He was "Aleck" at home, but to other people, from this time on, he was "Graham Bell."

Some years later Aleck made another change of name. Up to that time he had spelled his name "Aleck," but in one of his letters home when he was about twenty-five years of age, he signed himself "Alec" and in a postscript he adds, "Note the new spelling, 'Alec'."

CHAPTER FOUR

Grandpapa Bell

At this time, Aleck's grandfather, Alexander Bell, was living in London. As a young man he had been a shoemaker, as his forebears had been, in the University town of St. Andrew's, and like the cobbler in "Julius Caesar," he was "a mender of bad soles." But he was no ordinary cobbler, for he had distinct gifts as a comedian and possessed an exceptionally fine voice; and he knew his Caesar and Virgil and could recite "Macbeth" and "Hamlet" and "The Merchant of Venice" from end to end.

But one day a new cobbler sat on the shoemaker's bench; Alexander Bell had married and had gone to live in Edinburgh. There for awhile he played minor parts in the Theatre Royal. Then he returned to St. Andrew's and became a teacher of speech with a classroom in the University, after which he opened a boys' school in Dundee; and then, London! At first he gave private lessons, teaching young people to read and recite, what gestures to use, when to lower and when to raise the voice, how to bring out the vowel sounds—for what is known as "elocution" was a popular form of entertainment in his day. Alexander Bell himself gave readings from Shakespeare in public; and besides this, he gave public lectures on correct speech, how to treat defects such as stammering and faulty enunciation. He soon became well-known in London, and in time he became a professor of Elocution in the

University—with the cobbler's bench left far behind him—and he had the reputation of being the best reader and reciter of his day. He was a remarkable looking man, with an expressive face, firm but mobile mouth, large nose, the "Bell" nose, and a crown of snowy white hair that had once been jet black.

He lived in a quiet street opposite Harrington Gardens. He had occupied the house so many years that he seemed to be a fixture that could not change. But suddenly a great calamity befell him. His wife died, and he found himself alone, terribly alone, in the heart of the great city.

Another year dragged out its slow length. Then one day unexpectedly his son Melville (Professor Bell) came down from Edinburgh to visit him for two or three days, and among their many long talks they spoke of the boys.

"How is Aleck getting along at school?" Grandpapa asked—as the miller had done not so long before.

"Not very well," replied Professor Bell. "He's all right in mathematics, but he hates Latin, and he knows nothing of Greek; but he'll have to get it up. I don't know why any boy should hate 'Caesar.' The right kind of teacher should be able to make that story of the young standard-bearer live, and make a boy forget about its being Latin. 'Gerund-grinders,' old Tom Carlyle calls the dead-and-alive teachers of his day, 'who have no live coal within them.'"

"What are you going to make him," asked the old schoolmaster, "a teacher of speech?"

"I had always hoped so," said Professor Bell. "He should be our successor, Alexander Bell the Third, Professor of Elocution. But he's interested in too many things—collecting butterflies, and insects, and flowers—and won't settle down to work except at the piano. His ambition is to become a great pianist."

Grandpapa Bell had a far-away look in his eyes, recalling his own boyhood days. "No," he said as if thinking aloud, "that won't do. Aleck must learn Latin, and Greek too. If he is going to be a teacher of speech, he must know all about words. There is nothing like Latin to teach a boy to think clearly and express his thoughts concisely. And Aleck must begin at once to train his voice, and study how to read and recite. I wish I could help him." Then after a pause, "How would you like to send him down here to London and let me take him in hand for six months or a year. If he is willing to learn and will apply himself, I can help him. I'll put him through his paces. For a boy like Aleck a year in London is the best thing in the world—better for him than a year at school. At the end of a year he may know better what he wants to do. Let him come to me."

"His mother would miss him," replied Professor Bell, "and there's his music. But we can't stand in his way! It's an opportunity for the lad and he must take advantage of it."

CHAPTER FIVE

The Great Adventure

"LONDON?—AT GRANDPAPA'S? ME?"

Professor Bell had just told Aleck of his grandfather's offer, and Aleck could scarcely believe his ears.

"London!" he repeated, incredulously. "When am I to go?"

"As soon as you can get ready," replied Professor Bell. "You're a lucky boy. I wish I had the chance. Grandpapa is going to teach you. He has been a schoolmaster all his life, and you'll have to settle down to work for him."

Aleck could hardly take it in. London! London that every boy longs to see—fifty great cities rolled into one, and doubled twice over. Aleck had dreamed of it, and now his dream had come true! "London! London! London!" The name sang in his ears.

O London holds the hearts of men,
And London's paved with gold!

He would have to study Latin! No escape from that! But who would teach him? Grandpapa? Had Grandpapa a piano? What would he do all day long when Grandpapa was away? And Sunday? If only Melly and Ted could go with him!

"The boy is dreaming," said his father, "and no wonder! He's beginning to grow up. We'll see what London will do for him!"

Preparations were made for the journey and for the long winter in London; and to the boy it seemed as if time were standing still. But at length the appointed day came. It was a long journey for a fourteen-year-old boy to take, even if his father went with him, and Aleck was glad when it came to an end. It was getting dark when the train pulled into Euston Station. The lights, the noise, the crowds, the cries of station porters and cabbies were bewildering; but his grandfather pushed his way through the crowd. He engaged a hansom cab, and away they went, with Aleck's box beside the driver and the horse's hooves clanking sharply on the cobble-stones. Then, almost before they knew it, the cab drew up at Harrington Square. Then there was supper and bed, and a tinge of homesickness as he fell asleep. This was London, and for Aleck it was a great adventure, the first great adventure of his life!

It was not the habit of Grandfather Bell to put off the tasks that had to be faced; and as soon as Aleck's father had set out for home, he took the boy in hand. But before making a beginning with his studies he took Aleck to his tailor to have him outfitted with new clothes that were less rough and coarse than the loose tweeds that the boy had been wearing. The new tight fitting trousers made the boy utterly uncomfortable. Grandpapa insisted that he should be fitted out also with an Eton jacket and top hat. But what Aleck hated most was that he must carry a cane and wear gloves every time he went out. Across from the house at Harrington Square was a small park known as the Harrington Gardens, open only to the residents of the Square who had keys to it; and here Aleck was forced to parade for a certain time each day to satisfy the pride of the old schoolmaster. Aleck felt a genuine affection for his grandfather, but to the end of a long life, he recalled the misery which he had suffered from these new clothes during his first weeks in London.

Alexander Bell was methodical and thorough in the extreme; and Aleck found from the very first day, that his studies and his pleasures alike were planned to the minutest detail. Grandpapa was dismayed to find out how little the boy really knew. He had only a smattering of Latin and less Greek; and Grandpapa Bell had to begin at the first and drill the boy on his Latin grammar. The old schoolmaster took him in charge and taught him how to study. "Grandfather," wrote Aleck many years later, "made me realize that I was grossly ignorant of the ordinary subjects of study that every schoolboy should know. He made me ashamed of this ignorance and aroused in me the ambition to remedy my defects of education by personal study!"

When Grandfather Bell first thought of having Aleck come to London, the thought that he had in mind was that he might teach the boy to read and recite. "After all," he said to himself, "the boy was named after me. He is Alexander the Third, and he ought to be interested in the study of speech." Aleck, the old schoolmaster resolved, should be a teacher, like his father and his grandfather before him; and, in any case, practice in reading could do him no harm.

Back in the 1860's, in the days before the gramophone and the radio and television, and before the motor car was invented, reading and reciting were common forms of entertainment, not only at public entertainments, but at school concerts and church socials and in private homes. No programme was complete without recitations or dialogues; and young people took lessons in voice training and elocution so that they might acquit themselves creditably in public. To make a recitation "go over" with the audience, the performer studied what attitudes and gestures were most effective and paid special attention to inflection and enunciation and modulation of the voice.

Aleck had an unusually fine voice. He had learned to pronounce his vowel sounds, to watch for his "o's" and "a's" and "r's" and his father had taught him to enunciate clearly. Enunciation—that was the important thing, more important, even, than pronunciation. He had learned to speak in crisp clear tones.

When Grandpapa thought that Aleck was far enough advanced to profit by it, he took the boy into his own classes so that he might see and hear for himself how to deal with defects of speech and how to read in public; and Aleck listened once again to the old favourites which he had so often heard his father recite— "Horatius at the Bridge," that part about Astur, and the place where the bridge went splashing down into the Tiber; and "The Charge of the Light Brigade" (everyone was reciting that!); and there were stories in prose that he liked—"The Lark at the Diggings," and "The Christmas Carol" and the famous Trial Scene from "Warren Hastings," and half a hundred more.

And Shakespeare! Aleck knew some of the great speeches by heart. But there was one in particular that was a special favourite of Grandpapa's, "To be or not to be," when Hamlet was contemplating taking his own life.

> "To be, or not to be; that is the question;
> Whether 'tis nobler in the mind to suffer
> The slings and arrows of outrageous fortune,
> Or to take arms against a sea of troubles,
> And by opposing end them. To die; to sleep;
> No more—"

Grandfather had studied every word and phrase so as to give each line its proper shade of meaning; and when he recited it, Aleck was greatly moved.

"I want you to learn that," said Grandpapa. "You'll never be sorry, and someday you may be very glad."

"To be or not to be." Neither Aleck nor Grandpapa Bell could foresee that a day would come in Aleck's life when this question would be put in circumstances so very, very different from these.

It took Aleck some weeks to learn it well enough to please his taskmaster; but at length Grandpapa said, "That'll do just now, but we'll go over it again later on. You must recite it for your father when he comes to London again."

But there was much else in London that was an education for the boy, besides what his grandfather taught him. There were famous places to see, Westminster Abbey, the House of Commons, St. Paul's, Trafalgar Square, the Tower of London, and places where a boy could while away his time, such as the Zoo, and Madam Tussaud's, and the Horse Guards and half a score of others.

Aleck could scarcely have lived in London for a whole year under Grandpapa's roof without having been taken to hear Spurgeon, the most popular preacher of the day. The great tabernacle in which he preached seated six thousand people and Spurgeon had preached elsewhere to as many as ten thousand. Spurgeon was not yet twenty-eight. He had a clear and sympathetic voice which vibrated with emotion, and he moved his vast audience at will, from tears to laughter and from laughter to tears. One of the sermons that Grandpapa Bell and Aleck may have chanced to hear was entitled "The New Song and the Old Story." He took for his text, "O sing unto the Lord a new song," and his sermon was a simple and moving exhortation to his hearers to praise God in song. For Aleck, who was more interested in human speech than in anything else, there was one passage that might have seemed intended for him.

"The human voice, what a noble instrument it is! What a compass it has! Its low soft whispers, how they can hold us spellbound. Its full volume as it peals forth like thunder, how it can startle and produce dismay! Is not our tongue the glory of our frame? All the instruments that were ever devised by men are harsh and grating compared with the unparalleled sweetness of the human voice!"

This year which Aleck spent in London had a lasting influence on his whole life. "The year with my grandfather," he wrote years later, "converted me from an ignorant and careless boy into a rather studious youth, anxious to improve his educational standing by his own exertions and fit himself for college."

But this was not all. Life in the great city was an education in itself. In London he felt the heartbeat of the great world. Its surging thronging crowds, its busy life, its famous people and famous places, its romantic story, all left their impression. It was the turning point of his life, for while under his grandfather's roof he made, half-consciously, the choice of his life's work, and learned from "The Master" both what he should teach and how it should be taught. When he returned to Edinburgh at the close of the year, a new world was opening before him, and he was growing into young manhood.

CHAPTER SIX

"The Little Man"

WHEN THE TIME came for Aleck to return home, Professor Bell came down from Edinburgh to fetch him. He himself needed a rest and change, and he wanted to hear what Grandpapa had to say regarding Aleck's progress with his studies. Grandpapa was a strict and somewhat stern schoolmaster of the old type, but he reported that Aleck had improved greatly in his school work.

Before setting out on his return journey, Professor Bell took Aleck to see a new talking machine that his friend Sir Charles Wheatstone had brought back with him from Vienna. It was attracting a good deal of attention, and Professor Bell wished to see for himself what it was like.

"Can it really talk?" Aleck asked his father.

"No," said Professor Bell, "it can't talk, but if you wind it up and set it going, it can make a number of sounds that resemble words, but it just says the same thing over and over. Every once in a while someone comes along who claims that he has invented a machine that will speak. A few years ago one of these inventors made a machine that could talk and play cards, and add and subtract and say the multiplication tables. But it couldn't do division! And of course it was a cheat. There was a little dwarf hidden inside, who did all the talking!"

"But what about this machine?" said Aleck. "It does make sounds, doesn't it? What does it say? How does it do it?"

"I don't quite know," said Professor Bell. "I'll know better after I see it and hear it. In the inside, the part that makes the sound, there are tubes, and each tube makes a different sound. There is a tube for 'a,' another for 'o,' another for 'e,' and there are tubes for 'p' and 'm' and 'r'. If you put these sounds together in different ways, you get different words. Here are some of the words that the speaking machine can say: 'Opera, astronomy, Constantinople, Vous etes mon ami, Je vous aime, Venez avec moi, Augustus imperator.' If you study these words you will observe that the same sounds are used time and again in different combinations.

"There is some trick in all of these machines," said Professor Bell. "They can't really talk, for they can't think, and they can't ask questions or answer them. They can just make sounds, like a parrot; and tomorrow they'll just say the same words over again, that they said today—just like Perd. It's an automatic machine. All these machines are automatic."

"Automatic?" said Aleck. "What's that?"

"A machine that goes by itself, like an alarm clock, and does the same thing over and over as long as you keep it wound up. It's a 'self-actor'; that's what 'automatic' means. I wish I knew a little Greek."

"Humph!" said Aleck. "Latin's bad enough!"

"Over in Ireland," said Professor Bell, "a man once made a clock with a door in the dial, and every hour this door opened, and a little man came out and struck a bell with a hammer and said, in a squeaky voice, 'Past one,' 'Past two,' and so on, day and night the year round. It was an automatic machine."

"Where is it now?" interrupted Aleck. "I'd like to see it."

"The old man took it to pieces," replied Professor Bell. "He wasted so much time showing it to visitors that he couldn't get on with his own work."

When Aleck returned home he had so much to tell his brothers that he scarcely knew where to begin. But at length the opportunity came to tell them about the talking machine, and as Aleck told his story they plied him—and their father—with questions.

"Don't you think," said Melly, "that if we tried, we could make a little man that would talk, not just like the one you saw in London; but a little man with mouth and throat and lungs and vocal cords. I'd like to try. If anyone can do it, we ought to be able. But I wish I knew how that machine was made!"

"Down at the market," interrupted Edward Charles, Aleck's younger brother, "there's a Punch and Judy, and they can talk!"

"Huh!" said Aleck. "Punch and Judy! They don't talk. They're not even automatic. I watched the man pull the strings that make the lips move, and he does the talking. He just 'throws' his voice and keeps his own lips closed most of the time. He's a ventriloquist."

"A stomach-speaker," said Mr. Bell. "That's what 'ventriloquist' means. But what about the little man that you want to make? I wish you'd try it. If you do, when it is all finished I'll give a guinea to whichever one of you does the best work, even if the 'little man' can't say a word. But remember, you'll have to find out all about sound and know just how words are formed. Lips, mouth, and throat are the important things, and the 'little man' can't say anything unless he has a good pair of lungs and uses his tongue."

It was a terribly difficult thing for two such young lads to do; but they were eager to try. A guinea was a lot of money! They had a little workshop in the attic of the apartment house, and each had a corner of it. To begin with, they had to agree on which part of the work each of them would undertake to do. But there wasn't anything that Edward Charles could do—he was too young, and he had to be content with watching the older boys at work.

"I'd like to make the nose and mouth and lips and ears and maybe the tongue," said Aleck.

"That suits me," said Melly. "I'll make the throat and the lungs and that other thing—what do you call it?—the larynx—the thing that does the speaking. The lungs—a bellows'll do for that—there's one in our old organ. You can make a skull. Father's got one in his study. That'll be easy, but you're not going to win that guinea, not if I can help it!"

"We'll see about that," retorted Aleck. "Don't brag too soon. Slow and steady wins the race! That's what Grandpapa says."

"It isn't so easy as you think," said their father. "But you'll do your best with it. There mustn't be any guesswork. And don't keep running to me with every little difficulty you have. Try to work it out for yourselves, if you can. It must be your very own work."

It *was* a difficult piece of work for them, and if they had known how troublesome it was, they might never have started it. First of all, they had to study the different organs of speech: face, mouth, nose and throat, and the larynx, with windpipe and vocal cords; and they had to find out all about the materials they could use, and they had to do everything with the greatest patience and the greatest care and make all their measurements very accurate. Aleck had to make a study of the mouth and the different cavities and air passages connected with it. The lips had to be made so that they would move, but they must not be made of hard material. You could not do much talking if your mouth was made of wood. No, rubber was the only thing to use, and Aleck set to work to find out all about it. "Caoutchouc" was the dictionary name for it. What a word! Just like sneezing! Why not call it plain "rubber"? There were a good many things that the dictionary didn't tell about rubber, that were worth knowing. Does rubber, for instance, shrivel up or swell up, when you hold it close to the fire?

The most important thing that the boys had to learn about rubber was how to "vulcanize" it, to make it soft and pliable. Old Vulcan, who forged Jove's thunderbolts, didn't use rubber. He was a blacksmith, with a forge deep down under Mount Etna in Sicily, and he used fire and hammered them out on his anvil. The harder Jove's thunderbolts were, the better! But whether you vulcanized the rubber or not, it wouldn't do for the gums. The stuff they call "gutta percha" was better because it is more firm. It is the kind of gum (gutta) that comes from the "percha" tree.

Then there were the tongue and the cheeks and the soft part of the mouth and throat, which must not be too hard. The little man must have a tongue, but no tongue that Aleck could make could run as fast as his own. It could twist itself into a thousand shapes, and make, or help to make, a hundred different sounds. Besides, there are many different kinds of tongues. A toad has his tongue fastened at the opposite end from ours, so that he can stuff flies down his throat; but he has nothing at all to say, and he leaves it to the frogs to do all the croaking and all the piping from the ponds in Spring.

Aleck pondered for a long time over the problem of how to make the tongue, and he tried this thing and that. Then suddenly it came to him.

"I have it," he said to Melly. "If I use cotton batting, and put a thin covering of rubber over it, it should be soft and flexible. We're getting along!"

The bellows that formed the lungs were not hard to procure; and Melville succeeded in making an artificial larynx, or throat, of tin, with a flexible tube attached to it as windpipe.

In the meantime they were learning how words are made. "It's queer," said Aleck, "how they come tumbling out of your throat, and how you can fit them together to say just what you mean."

Sometimes the boys became impatient because they seemed to be making such slow progress. It occupied their spare time for months, and they couldn't think or talk of anything else. Time and again, each of them had to help the other, and they had to consult their father about their difficulties in spite of what he had said to the contrary.

But at length the "Little Man" was complete, as perfect as they could make it, and nothing remained but to fit together the parts that each had made. As the time grew near they were all excited, and they could scarcely contain themselves. Then came the real test! The parts were fitted into place; the bellows were set in motion, and the air came through with a harsh rasping, gurgling, strangling, whistling, hawking sound, as if half a dozen distant alarm clocks were going off at the same time. Then Aleck opened the lips and set the tongue in a different position, so that when the air was forced through, the "Little Man" said "M-ah! M-ah!" which sounded like "Mamma." Then they made some slight changes and tried it again. But no! It would make all sorts of queer noises, but it wouldn't *talk*.

"I know why," said Aleck, "but I can't do anything about it. The 'Little Man' has no muscles, and he can't open and close his throat, or touch his teeth or the roof of his mouth with his tongue, or puff out his cheeks or press his lips together. It's just like blowing through a tin whistle. He can make sounds but he can't make words."

The boys were terribly disappointed and were ready to weep with vexation. All their work seemed to them to have gone for nothing; for in spite of what their father had said, they had expected the "Little Man" to say something; but it couldn't, or wouldn't say anything but" Ma-Ma." Then something happened that made them laugh.

Their house was a flat, and a common stair led from the basement and the street door up to the landings of the upper floors. It was a stifling hot day; and when they left the door of their attic workshop open, the cries of" Ma-ma" rang through the house as if a child were in agony. Then after an hour or two of these torturing noises from above, a woman from the floor below opened her door, and looking up the stairway said in an exasperated voice, "It's a wonder they can't keep that baby quiet!" The two boys laughed and shouted and danced about the "Little Man," the machine that tried to talk without help, but couldn't. And the woman in the flat below wondered what all the commotion on the top floor was about. It was all very trying! You might as well have a parrot or a Punch and Judy show in the house as a "Little Man" with a wooden brain, that hadn't learned to talk!

But although it seemed to the boys that their experiment was an out-and-out failure, their father was delighted with what they had accomplished.

"If the 'Little Man' could only speak," he said to them, "he'd shout 'Well done!' but he can't speak a word and I'll have to say it for him. 'Well done, both of you. I'm proud of you.'"

But there was one thing that troubled him. He had promised a guinea to whichever boy should do the better work, but they had both done so well that he couldn't make any decision. Melville was the better workman of the two. Aleck was always clumsy with his fingers, even when a grown man. But he designed his work better than Melville, and he supplied the ideas which Melville put into practice. In the end, Professor Bell did, no doubt, what any other father would have done—he divided the money between them—and all three were satisfied; and the "Little Man" didn't have a single word to say!

CHAPTER SEVEN

"Visible Speech"

During these years Professor Bell lived a very busy life. He gave readings in public; he conducted classes in Elocution; and he spent long hours each day in giving instruction and help to boys and girls, and men and women too, who were suffering from stammering and stuttering and other defects in speech. Neither he nor anyone else had thus far found an easy way to remedy these speech defects. But one day as he was turning the problem over in his mind for the thousandth time, a way of meeting the difficulty suddenly occurred to him, as if it were a flash of inspiration. "The reason," he said to himself, "why these unfortunate people suffer from speech defects is that they have not learned the correct position of the lips, tongue, and other organs of speech. But might it not be possible to show them the correct positions in drawings or pictures? Might not drawings or diagrams be made to show the action of the vocal organs in uttering different sounds?" And he set to work to make a series of skeleton pictures, each of which stood for a certain position or movement of the lips and tongue. It was a special kind of alphabet. Here at last was the very thing he had so often dreamed of but had never been able to put into readable form.

As he worked his idea out it occurred to him that besides helping to correct imperfect speech such as stammering, these "pictures" could be used in the teaching of foreign languages. Any language might be written in the new symbols; and anyone who could consult these symbols could then pronounce that language.

When Professor Bell had completed his "alphabet" he had to find a name for his system, and after trying out a number of others, he decided to call it "Visible Speech." That was easy to understand, and since speech is something to hear, and not something to see, the name "Visible Speech" catches your interest at once. You can, of course, see people speaking, moving the lips and tongue, but you cannot see what goes on *inside*—in the vocal cords—or still further inside, within the brain where words are chosen and shaped and set to music and woven into whatever pattern you may design.

When Professor Bell announced his invention of Visible Speech, he had to explain it to the general public and demonstrate it to the smaller groups who were teachers of speech. At first there were only twelve of these "pictures" or symbols in the "Visible Speech alphabet," and they had to be thoroughly studied and learned, and the teacher of Visible Speech had to know how to join them together to make new combinations. Once the "alphabet" was mastered, you could write down in "Visible Speech" language all kinds of words and all sorts of noises—the cries of the "rags, bones and bottles" man, the barking of the dog that kept people awake all night, the clanging of the fire-bell and even the barrel organ that played all up and down your street of a summer afternoon. That was a hard one to do!

Of course there were people who found fault with "Visible Speech." "It is too difficult to learn," said one. "There are all sorts of sounds," said others, "that can't be written down, the song of the nightingale, for instance."

"No," said Professor Bell, "you're wrong. I can write down the symbols for any noise you make, and if you doubt it I will be glad if you will put me —and the system of 'Visible Speech'—to the test. Nothing will please me better than to have you try it out."

And so it came about that on a certain September afternoon at the home of Mr. Macrea, their clergyman, Professor Bell undertook to demonstrate to a little company of friends that "Visible Speech" could actually be put into practice. But in order to do so he had to have the help of the two older boys. They had learned the "Visible Speech" alphabet so thoroughly that they could not only write sentences in "Visible Speech" but could read what the symbols said. They were all eagerness, for they wanted to show just what a fine thing "Visible Speech" was and how clever they themselves were to be able to read it. When the afternoon came—and with it the members of the "jury" both the boys and their father were a little nervous lest anything might go wrong.

When all was ready for the test, the boys were sent out of the room, beyond earshot. Then the company, each in turn, made whatever sounds they pleased, while Professor Bell translated them into signs and wrote them down. There was the word "Welcome" in Gaelic, and "The Campbells are coming," as if droned on the bagpipes, and the cackling of a hen, and the first line of the Aeneid, and one of the company yawned loudly—at which they all laughed, and Professor Bell wrote that down, too, in the language of "Visible Speech." Let the boys try that! Then the two lads were given some minutes to study their father's drawings and one after another they reproduced the original sounds. Mere boys though they were, they proved beyond question that the system would work.

But if any of the neighbours had looked in at the window on that peaceful Sunday afternoon, or had listened at the keyhole to these outlandish noises followed by gusts of laughter, they must have thought that this quiet dwelling had been transformed into a mad-house, and that these ordinarily sane and sensible Edinburgh people suddenly had lost their reason.

Not very long after this, the system of "Visible Speech" was put to an even more severe test, when a demonstration of how it works was given before a large audience in a public hall in the city. On this occasion, Aleck acted as his father's assistant, all by himself, and translated the "Visible Speech" symbols back into their original sounds without a mistake. It was a test of Professor Bell and the boy, just as much as a test of the system. Just once in the course of the evening, Aleck came near to being puzzled. He looked at the paper with the symbols which his father had written down, and at first could make nothing of it. Surely the symbols must be wrong! But he went through it as the paper directed, and a great burst of laughter greeted the sound. It was an imitation of a man with a buck-saw, sawing wood, a sound that is most unmusical, but unmistakable.

When Aleck followed the directions of another peculiar group of symbols that was handed to him, the sound that he made was one he had never heard before. The test had been given by a professor of Hindustani who was much pleased with the result, for the sound was of a certain Sanscrit "T" which English-speaking students found very hard to master.

The success of these tests led Professor Bell to predict that his "Visible Speech" could be used in teaching deaf children to speak. Since such children could not learn by hearing and imitating others they might learn by seeing how words are pronounced. And, in later years, "Visible Speech" was so used.

Meantime, all his work with his father attracted young Aleck's attention to problems pertaining to speech, and without his knowing it, prepared the way for what was to be his life-work.

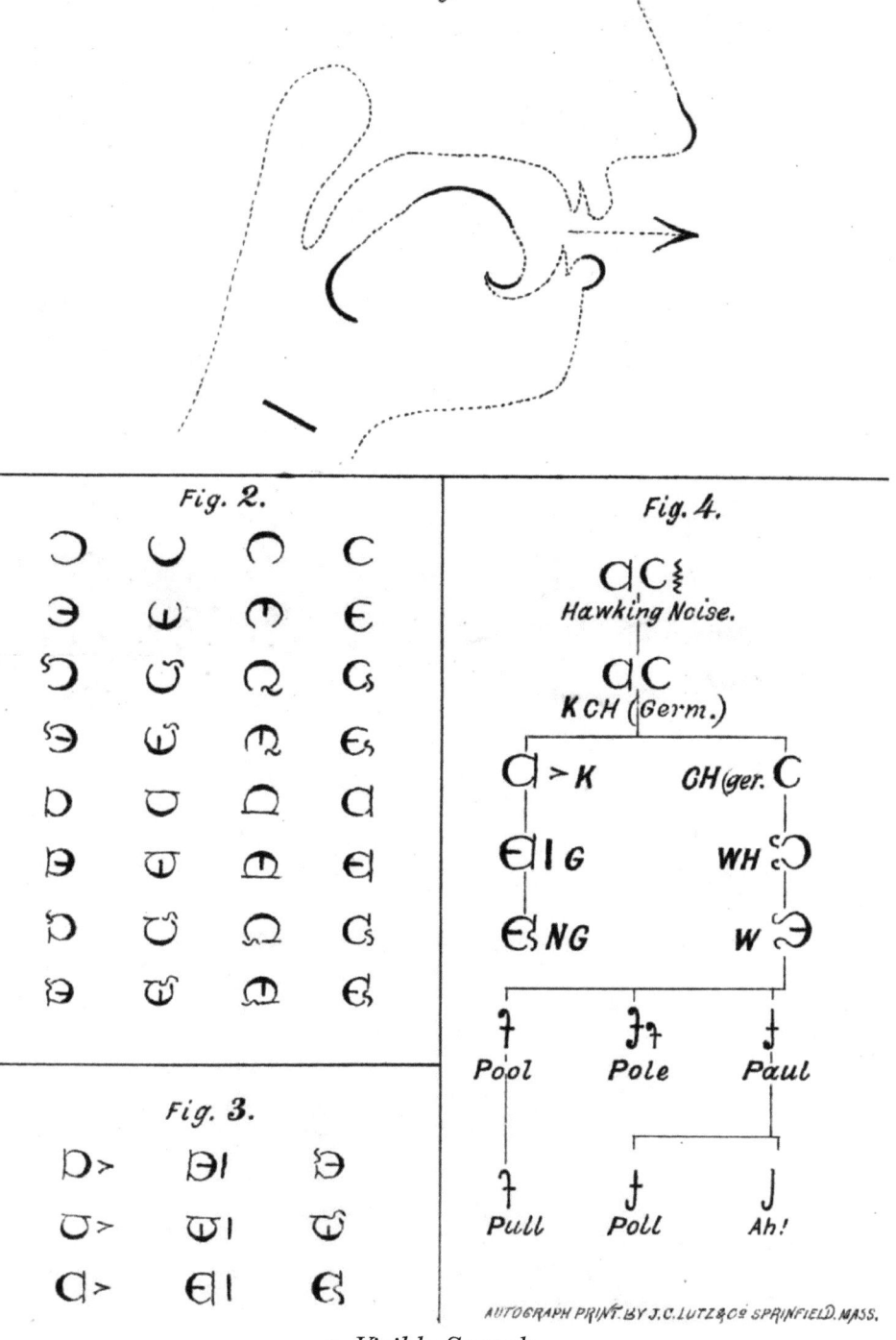

Visible Speech

Early advertisement of the telegraph 1845.

The first Morse telegraph 1835.
Above the transmitter.   *Below* the receiver

Part 2.

THE SHADOW.

CHAPTER EIGHT

A Career

WHEN ALECK WAS in London his father supplied him with some spending money, but not too much. A boy in London would need some money, of course, a shilling a week, it might be. But Grandpapa Bell paid the small sums that were necessary for admission to the Zoo and Madame Tussaud's and the Tower and other places like that, and he paid Aleck's train fare and journeys by coach. But when Aleck had returned to Edinburgh his father could see no reason for keeping up his extravagance of so-called "pocket money." What could a boy in Edinburgh need of anything more than his "keep"? Mutton pies and Chelsea buns and other dainties displayed in the shop windows might tempt a boy, but he was better without them. And if Aleck were given spending money, what about the other two boys?

Aleck did not protest. But he was growing up, halfway between boyhood and youth, and there were things that he coveted, and places that he longed to explore. These things cost money but his pockets were empty! He wanted to earn some money of his own, no matter how little, and he was getting old enough now to know how to spend it. His father did not understand, and he could not talk it over with him. But there was Melly! *He* understood. He too would like to earn money of his very own—but what could they turn to that would bring in even a sixpence?

"Well," said Aleck, thinking of what he himself might do, "we could give readings from Shakespeare. 'To be or not to be,' I know that, and 'The quality of mercy' and that speech about St. Crispin's Day." He struck an attitude and began:

"This day is call'd the feast of Crispian.
He that outlives this day and comes safe home,
Will stand a tip-toe when this day is named,
And rouse him at the name of Crispian.
He that shall live this day, and see old age,
Will yearly on the vigil feast his neighbours,
And say, 'To-morrow is Saint Crispin's.'
Then will he strip his sleeve and show his scars,
And say, 'These wounds I had on Crispin's day.'"

"No," said Melly interrupting him. "That wouldn't do. People won't pay money to listen to us. With Papa it's different, that's his profession!"

"Well, then," said Aleck. "I'll teach. There's always some fellow who wants a tutor!"

"Yes," said Melly, with a laugh. "You'll teach Latin! You'd make a fortune at that!"

Aleck made a wry face. There's nothing that brings a boy face to face with the realities of life more quickly than having to earn his living. "No," he agreed, a little sheepishly, "that won't do. Let us look at the advertisements in 'The Scotsman.' There may be something that'll suit us."

Day by day for a whole month, they scanned the closely printed "Want Ads" in their father's copy of "The Scotsman," when he was not there to question them. There were advertisements for nurse-maids and journeymen printers, and gardeners, and coachmen and the like, but no one wanted a fifteen-year-old lad who could teach Elocution, or thought he would like to try! The School Committees didn't know what opportunities they were missing! And then, just as they were about to give up, there appeared an advertisement for two pupil teachers, who could teach Elocution and Music in a boys' school, Weston House Academy, at Elgin in Morayshire. The "honorarium" ("That means 'salary'," said Aleck) would be small, but the pupil-teachers would have some time for private study! "Wanted, two pupil teachers." Where was Morayshire? It had to be looked up! The boys were excited; this was their opportunity!

> "There is a tide in the affairs of men
> Which, taken at the flood, leads on to Fortune;
> Omitted, all the voyages of their life
> Is bound in shallows and in miseries."

That was how it seemed to the boys. They read the advertisement twenty times over.

"I can teach," said Aleck. "I've watched Grandpapa, and Papa too. I can teach 'Visible Speech,' and I can teach Music just the way Signor Bertini taught me, and if I have to, I can teach Arithmetic."

"And Geography?" interrupted Melville, with just a hint of sarcasm.

"Yes," said Aleck. "I can teach Geography too, if I *have* to. But I don't like it. Where's Morayshire?"

They didn't tell their father that they were going to answer this advertisement—he would have forbidden it, of course. But Aleck wrote out an application in a cramped boyish hand, giving their names and offering to teach Elocution and Music "by the most approved methods." But they didn't state their ages. That would have spoiled it all. Aleck was nearly sixteen and Melville just two years older, but both were tall for their age.

"One thing more," said Aleck doubtfully, "we have to give references. The advertisement says so. What'll we do about that?"

"There's Papa," said Melville. "He'll do. Give them his name. Everyone knows *him*."

When Mr. Skinner, the Headmaster at Weston House, received this application, he was puzzled. "Who are these young men?" he said to himself, 'Graham Bell,' and 'Melville Bell.' That can't be our Melville Bell! It must be a mere coincidence. I'll have to get to the bottom of this!"

So he wrote to Alexander Melville Bell and made inquiries. Are these young men your sons or your nephews? How old are they? Can they really teach?

When Professor Bell received Mr. Skinner's letter of inquiry he was annoyed. He called the two lads in. "What do you mean by playing a trick like this?" he said. "What'll Mr. Skinner think of us? You ought to know better! Teach Music and Elocution? The idea! I can't let you go, Melly, I need your help, and you haven't been looking well lately. But what about you, Aleck? Do you think you could teach?"

"I could try, sir," said Aleck hopefully. His father was toying with the idea, he could see that. "I really want to earn some money."

Mr. Bell looked at Aleck. He was not quite sixteen—but a clever boy! A lad to be proud of! Mr. Skinner might do worse! And it would do the boy good—bring him down to earth. He was good tempered and amiable, a general favourite—and he would be earning some money.

"I'll think it over," he said to Aleck. "We'll talk about it to-morrow!"

In the meantime he wrote to Mr. Skinner and told him the

whole situation, and after an exchange of letters, Aleck was engaged as a pupil teacher, to give instruction in both music and Elocution, at £25 for the year, with his maintenance! Aleck was in the seventh heaven! The days could not move fast enough. He hated to give up his lessons and to leave his father and mother and brothers; but when the time came, he set out for Elgin feeling very grown up and very proud —and just a little homesick.

But his mother's letters, recounting little everyday incidents, kept him close to the beloved home circle.

And, besides the "news" of home, the letters to Aleck were full of motherly admonitions about his health which was causing Mama some anxiety. His eyes were troubling him and he suffered severe head aches. The understanding Mama wrote, "Teaching Music and Singing is one of the most irritating branches of tuition that can be and is enough of itself to produce nervousness." These were the earliest of those warm, stimulating letters which went to him from his mother whenever they were separated—letters which he preserved to the end of his life.

His work at Weston House was tedious and monotonous; but in teaching small boys the rudiments of Elocution and Music, he had to make a study of sound; and in so doing he found out that the human voice is not so simple a thing after all, but that the muscles of the throat, and the vocal cords, and the cavities of the mouth, and the tongue, and hollow bones, all have a part to play in producing vocal sounds. He soon became interested in trying to analyse the human voice, and before long he began to make experiments with tuning-forks to discover, if possible, what different vibrations enter into the tones of the voice, and how the vowel sounds differ, one from the other. This was a new interest for him. He was just on the threshold of the vast new world of Science; and the study of resonance (sounds prolonged by vibrations) became an absorbing passion for the sixteen-year-old youth. He soon realized that until he had mastered everything there was to know about vibration and resonance he could not go much further in his studies.

But, boy though he was, he already knew that the human voice is the most wonderful instrument in the world. It was the lyre that Pericles and Demosthenes and Cicero and the Apostle Paul had used. Chatham and Burke and Sheridan, and Pitt the Younger, and Lincoln, the rail splitter, and Gladstone in his own day had played upon it and had moved men to joy and laughter and madness and tears.

CHAPTER NINE

The Enchanted Lyre

At the end of a year Aleck resigned his position at Weston House and went to the University in Edinburgh to study Latin and Greek. His mother, in her letters, had been urging him to give up the attempt to teach and study at the same time and to yield to his father's wish that he continue his studies at home. Aleck reluctantly returned, but when he was seventeen he went back to his position in Morayshire for another twelve months.

In the following year, 1865, Grandpapa Bell died at his home in London—sturdy, honest, enterprising Alexander Bell. Now that he was gone, the work that he had undertaken had to be carried on and provision had to be made for teaching his private pupils and his classes; and Aleck's father had to decide just what he should do. Should he remain in Edinburgh, where he was already well known? Or would it be better for him to leave his classes and his other engagements in charge of his son Melville, and go to London to continue and extend Alexander Bell's work? He was still in his forties, in the very prime of life. He had built up a reputation as a teacher of Elocution and had become famous as the inventor of Visible Speech. How could he hesitate? London seemed ready to welcome him, and the door of opportunity was thrown open wide. All that he longed for, all that he strove for, was there —worldly success, friends, fame, everything that his eager spirit could desire. London drew him as with a magnet.

And what of Aleck? He felt his grandfather's death keenly, but for the time being, he decided to remain in his old position in Morayshire. He was as eager and active as ever. Aside from his work in the classroom, he was wholly absorbed in the study of sound and he was wildly enthusiastic over every new experience or adventure. This business of resonance, which had so much to do with the complex quality of the human voice baffled him, and he still carried on his experiments with tuning forks. But there was no one at Weston House with whom he could share his enthusiasms or who could answer his questions. He wrote long letters to his father, but Professor Bell knew very little about the new science of "physics" as applied to the human voice. He was interested in speech and in Elocution and not in mere sound.

But when Aleck came to London on a visit, his father introduced him to two of his most intimate friends who were authorities on the subject of sound. One of these was the distinguished philologist ("student of words") Alexander John Ellis. When Aleck told him about the experiments that he had carried on at Elgin, he was greatly interested in the lad. He invited Aleck to his home and talked to him about the great German scientist Helmholtz, whose works he had translated into English. This opened up a new field to the boy, and he set himself to work to repeat the experiments of Helmholtz for himself. This was the beginning of his study of electricity. It did not lead to anything for months to come, but the seed was sown and ready to germinate when the time came.

A few days after his visit to Mr. Ellis, Aleck's father said to him, "I'm not able to give you much help with your experiments, but I wish you could have a talk with Sir Charles Wheatstone.

He knows more about sound than anyone I know of, except Mr. Ellis, and he will help you if anyone can."

"Wheatstone?" replied Aleck. "You took me to see the talking-machine that he brought from Vienna."

"Yes," said Professor Bell. "He is a very celebrated man, but he is very modest and talks very simply. In his younger days he kept a shop where he sold musical instruments. That is how he came to make a study of sound. Then he became interested in electricity; and his discoveries and inventions have made him famous."

"He is a very busy man," continued Professor Bell, "and I don't know whether he will be able to spare time to talk over your problems. But I'll try to arrange it."

Aleck sat silent for a moment, then took up the conversation again.

"Didn't he have something to do with the Atlantic cable?"

"That's why he was knighted," replied Professor Bell. "If it had not been for his experiments with the underseas telegraph, the cable might never have been laid."

"But it has been laid," said Aleck, "and someday we'll be able to *talk* by telegraph."

Professor Bell laughed. "You'd better ask Sir Charles about that. He knows everything there is to know about such things. No one would have dreamed, half a century ago, of sending messages under the sea by telegraph, two thousand miles or more."

"Dots and dashes," said Aleck, "sending signals by stopping the current and starting it again! If only we could make the current carry sounds, words, real live words, that *would* be something worth while!"

"Dreaming!" said Professor Bell. "We'll never see it in our day. It's impossible, quite impossible!"

When Professor Bell spoke to Sir Charles about Aleck's interest in the study of sound, and told him of the experiments that he was carrying on, no one could have shown a greater interest in the lad or have been more helpful to him. He gave up a whole evening to him. He answered Aleck's questions and discussed his experiments; and then he spoke of some of his own inventions as if they were all very simple things. One of these was a contrivance known as "The Enchanted Lyre." It consisted of two musical instruments. One of these, a piano or an organ, was placed in an upper room, and from the ceiling of the room directly below was suspended the second instrument, a musical box shaped like a lyre. The two instruments were connected by a rod or tube which was concealed from sight. When the piano in the upper room was played, the vibration was carried down to the lyre, and to the casual visitor it seemed as if the lyre were being played by unseen hands and was enchanted.

Aleck was interested in this ingenious "toy" in which music was carried over a rod or a wire. It was, of course, a mechanical device, and Sir Charles never succeeded in making the wire "talk." But he spoke to Aleck about the possibility of such a thing, and as time went on, the dream of Electric Speech began to take form in the boy's mind. These talks with Sir Charles and with Mr. Ellis provided the stimulus that he needed and he returned to his classes and his experiments with new interest and new eagerness.

CHAPTER TEN

The Alarm

During the two years immediately following the death of Grandpapa Bell, the family became adjusted to new conditions and new surroundings. Professor Bell took up his new duties in London as a teacher of Speech, and became Professor of Elocution in the University of London. Melly, now happily married, took over the work in Edinburgh where his father had left off, while Aleck completed another year of teaching in the Academy at Elgin in Morayshire. But now with the death of Grandpapa Bell still fresh in mind, another blow fell upon the Bell household. Edward Charles, the youngest of the three boys, just nineteen years of age, was stricken with "consumption," and died. We call it "tuberculosis" now, but "consumption" is an expressive word which means "a wasting away," and the very name brought with it a shudder of horror. Aleck's heart went out to his father and mother in their grief. At the close of the year he gave up his position at Elgin so that he might be nearer to his parents; and he took another teaching position at Bath, which was only a hundred miles from London.

Bath was a ghost town, with its old Roman baths, its great Pump Room, its ancient Abbey and its old stone houses looking down from the amphitheatre of hills around the town. It lived on its past, and ghostly figures walked its streets, Beau Nash, Pope,

Garrick, Dr. Johnson, and Pitt the Elder, and Gainsborough, and Lord Nelson, and Wolfe setting out to join his regiment at Quebec, and throngs of famous men and women. And hither, to the old grey city, its voices silenced now, and its gay life a memory of the past, came young Aleck Bell to Somerset College, to teach the art of speech to a new generation in this new age.

But Aleck's parents still disapproved of his independence; and his mother wrote, "Young birds are very prone to try the strength of their wings too soon. The parent birds know best the proper time for independent flying. You should not have your mind distracted by the care of earning your bread." She was interested in his experiments and wrote about the discussions going on in Parliament regarding "telegraphic communication with foreign countries" and she sent him a copy of a new publication, "The World of Science." But her chief concern was about his health and her letters were full of advice about his food, his not sleeping well, his need of exercise, and above all the necessity of wearing his flannels until warm weather, lest he should take cold. After the loss of Edward, she wanted Aleck under her wing. Aleck protested that he disliked London, where the family were now living; she retorted, "Your dislike of London should not weigh for a moment. You will like it just as well as Edinburgh after a little while." And so, at the end of a year, Aleck left Bath and went to London to take charge of some of his father's classes, and incidentally to take lessons in Anatomy at the University.

It was a busy life, and as time went on it became for Aleck, if possible, even more busy. As the author of Visible Speech, Professor Melville Bell had become almost as well-known in America as in England, and early in 1869 he received an invitation to give a series of lectures in Boston. In the old myth, Atlas, who bore the weight of the world on his shoulders, asked young Hercules to bear the burden long enough to give him a moment's ease; and the unsuspecting youth gladly took the great Globe on his own shoulders. In old London during these few months young Aleck played the part of Hercules to his father's Atlas, and undertook to fill his father's engagements as well as his own. The lectures in Boston were a crowning success, and Professor Bell made friends everywhere. But at the very moment of his triumph a dark shadow fell across his path. Melville the younger, now happily married and settled in Edinburgh, fell a victim to consumption. Aleck cancelled his engagements and hurried north to render what assistance he might; but it was already too late. Melville James, like a flower plucked in the heat of the day, faded away and was no more; and Aleck, bereft of the playmate of his childhood days, was disconsolate.

Professor Bell hastened home from Boston with the plaudits of New England in his ears, only to realize how empty and transitory a thing is fame. It seemed to him for a moment that he had nothing more to live for. Of his three sons only one was left alive. A blind terror gripped the parents' hearts, for it was quite evident that Aleck also bore the marks of the same dread disease upon him, the same deathly pallor, the same fits of weakness, and the flush of recurring fever. To Professor Bell there seemed to be only one course of action possible. London with its fogs and its contagions and its unending slavery—London, the old Siren, was no place for a youth of twenty and over, with the hand of death upon him. The doctors shook their heads and advised a change to a drier, milder climate. It seemed to be the only hope, and Aleck's parents weighed anxiously the advantages of this or that part of the new world in which they might possibly escape from the impending disaster.

As a young man, Professor Bell had himself spent four years

in Newfoundland—a clerk in a business house—when his health was threatened; and now in this crisis he thought of making his home there. But the Bells had friends in Canada, a family named Henderson, in the little town of Paris in Ontario. They wrote to Mr. Henderson, and a month later they received his reply. "This is the best climate in the world," said Mr. Henderson. "True, the winters are cold, but the air is dry compared with that of England. I have in mind a small property on the Grand River just outside of Brantford, which is an ideal location. It is on the heights overlooking the Grand River Valley and I think it would suit you." This was a strong recommendation, and the balance dipped in favour of the milder summers and colder winters of Southern Ontario.

The decision to risk everything in the hope of finding a better climate called for courage. The blow fell even more heavily upon the father than upon the son. He had earned a name for himself. He was at the very height of a successful career; and for the first time in his life he was in easy circumstances. He had made friends. His position as professor in the University of London was secure. But to carry out his decision, on the physician's orders, all this had to be sacrificed.

And what of Aleck's mother? She must exchange the comforts and enjoyments of the greatest city in the world for the hardships of a life in the wilderness. But the spirited, courageous lady did not hesitate. To them there was no alternative. The life of her son was in danger, and there could be not a moment's hesitation, for what would life be worth if they were bereft of this only remaining son? They took immediate steps to carry out their resolve. Within a month's time their effects were disposed of and arrangements were made for the voyage to Canada. Farewells were taken, and with heavy, foreboding hearts they set out to begin life over again in the new world.

For Aleck himself the change was no less momentous. He was leaving home and friends. After months of ceaseless activity, he was compelled to rest. The sea voyage was a wholly new experience. He spent long hours on deck resting, watching the swell of the waves and the gulls that followed the ship.

The laying of the Atlantic cable only a year or two before was still fresh in his mind as he looked out over the grey sea, and

he marvelled at the courage and decision of the man who had planned and accomplished this great feat. He could remember as if it were yesterday, when the first cable was laid, the year after Grandpapa Bell died, and now, as he closed his eyes drowsily, it seemed to him as if he could hear the whisper of words, millions of words, like the flutter of soft wings over the great ocean floor beneath.

"The sea! the sea! the open sea!
The blue, the fresh, the ever free!
Without a mark, without a bound,
It runneth the earth's wide regions' round.
It plays with the clouds; it mocks the skies;
Or like a cradled creature lies."

And the sonorous lines that every schoolboy knows:

"Roll on, thou deep and dark blue ocean, roll!
Ten thousand fleets sweep over thee in vain!
Man marks the earth with ruin; his control
Stops with the shore."

The time passed slowly, and it seemed to him as if the voyage might be endless. It was a thrilling moment when he first saw on the far horizon the shadowy outline of what might prove to be land. Then, entering the Straits, they sailed for two more days across the great Gulf, and "under looming shores," and saw the misty headlands of Labrador one by one fade from sight. And then the long day's voyage up the majestic river, past the mouth of the Saguenay, past Riviere du Loup, and past the Falls of Montmorenci and the Island of Orleans, to the frowning rock fortress of Quebec. It was an experience never to be forgotten. The immensity and majesty of the panorama moved him as nothing had ever moved him before, and at the same time it exhilarated him and made it hard for him, impulsive and active as he was, to remain still. But in some curious way it made him feel how insignificant a human being is, a mere speck, an atom in the infinite universe of space.

CHAPTER ELEVEN

Tutela Heights

At Quebec there was the usual long delay, the tedious disembarkation, and the inspection of baggage by Customs officers. Then the train with its uncomfortable cars, draughty, dusty, rackety, fatiguing, with noisy puffing engine and raucous whistle. It made them homesick for the comforts of old-world travel which they had left behind. But the countryside was new and different. From the train windows they looked out on quaint farms and village homes with steep roofs and white palings; and country roads with slow oxen hauling farm carts piled high with hay; and here and there they caught glimpses of the great river. At Montreal they saw very little of the city itself.

For them it was merely a place for changing trains. Then up and away, and out to the open country once more, past Ontario farms, fields rough and unkempt, and endless rail fences wriggling their way through clumps of golden-rod and asters; and wayside villages half-asleep in the late summer sun. What a world for a young man who had never in his life been out of England and Scotland! Stone-piles in the midst of brown pastures; golden grain fields ready for the harvest; grassy country roads; and sluggish streams winding through valley lands dotted with drooping elms. A world in the rough, and still in the making.

And now, at last, Toronto, in those days a quiet and rather

sleepy city of some thirty thousand souls. They were nearing the end of their long and tiresome journey, two more short laps only, from Toronto to Hamilton, and out over the long steep grade to Brantford and the little town of Paris a few miles farther west. Professor Bell's friend, Mr. Henderson, was at the station there, to meet them with a heart-warming welcome.

The next few days passed quickly. Paris itself was a picturesque little hamlet, nestling in an elbow of the Grand River, and famous for its plaster-of-Paris and gypsum. Mr. Henderson had arranged that the people of Paris should have the privilege of hearing Professor Bell read and recite; and on another afternoon Professor Bell explained his system of Visible Speech for the benefit of the teachers and parents.

But both Mr. Henderson and the Bells were anxious that they should get settled as soon as possible; and on the very next day after their arrival, the two older men drove to Brantford to have a look at the property that Mr. Henderson had mentioned in his letter. It was situated about two miles from the centre of the town, eight miles distant from Paris, on the west bank of the Grand River, and on the high ground known as Tutela Heights. Brantford was then only a small town, or overgrown village, of six or seven thousand inhabitants, lying out on the higher ground overlooking the broad valley of the Grand with high banks on either side.

The property on Tutela Heights consisted of ten and a half acres, an orchard and garden, pasturage for a cow and horse, a good frame house and a small barn; and some acres to spare for whatever they chose to plant. But for the present it wasn't so much the farm and garden that attracted Professor Bell as the peacefulness of the scene and the clean pure air that would help to restore the invalid to health and strength once again.

Behind the farm, at the foot of a precipitous bank ran the Grand River, and through a veil of trees along the edge of the cliff, the new arrivals might catch a glimpse of chimneys of tall factories, for Brantford was, even at that time, making claim to being a wide-awake manufacturing town. To Aleck it was all fresh and new. The river fascinated him, for it was never the same for two days in succession, and most of the time it lived up to its old name, "The Rapid River," which the early French explorers gave to it. It came down from the great swamps in the North country, a hundred miles away, and in Spring and Autumn when it was in flood it carried all before it. It was an ideal spot for a convalescent, especially for one who wished for time to dream and plan; and in the garden and orchard there was enough light work to prevent the young man from becoming impatient at the slowness of recovery.

Neither Professor Bell nor Aleck knew the first thing about gardening or farming; and they had to learn the simplest things about Indian corn and cabbages and onions, and other things. But a few hundred yards away lived the McIntyre family, the best of neighbours, and, a little farther down the road, Mr. and Mrs. Brooks and their son, who proved to be friends in need; and they were called upon whenever the newcomers found themselves in difficulties.

It was impossible for the Bell family to get along without a horse and buggy, and they had to learn how to care for the horse and how to harness it. It was one thing to drive straight ahead along a quiet country road where they would meet no one, and another thing to turn around on a busy street and keep out of the way of other vehicles; for even though there were no motor cars and no bicycles, it was a busy little town and the downtown streets were none too wide. But with the buggy, and in winter the sleigh, they were able to drive over to Paris to visit their friends the Hendersons, or to follow the twists and turns in the road down to the next big "bow," or bend in the river, where the Honourable George Brown made his home.

Within two or three months Aleck's health began to improve. When the Bells left London to settle in Canada his physicians were doubtful of his recovery and gave him only a few months to live. But rest, open air, and sunshine helped to check the disease, and he began to take a renewed interest in the study of the human voice, which he had been forced to give up on account of his illness.

The old homestead was an ideal place for his experiments; few people passed along this quiet country road; and his mother was so deaf that the noises made by Aleck with piano and tuning-forks, and his own voice, all day long, didn't disturb her. But the neighbours and chance visitors who heard him "oh-ing" and "ah-ing" and making all sorts of weird noises at the piano, or caught glimpses of him making faces in the mirror, with wide open mouth, as they approached the house, shook their heads and tapped their foreheads, and said among themselves, "What a pity! He seems such a nice young man, but he is a little 'queer.' It must be a great disappointment to his parents."

During these months of idleness Aleck was becoming more and more interested in a very important problem relating to sound. The invention of the telegraph thirty years before had meant an untold saving of time in the sending of messages to distant points, but thus far it had been found possible to send only one message over a wire at a time. Young Mr. Bell knew, as did a hundred others, that a fortune awaited the Tutela Heights inventor who could devise some means of sending several messages over the same wire at one time without jumbling them. Ever since his talk with Mr. Ellis on his visit to London he had been interested in electricity, and he tried to repeat the experiments of Helmholtz, the great German scientist. But the loss of his two brothers, followed closely by his own serious illness, had broken in upon his studies and interrupted his experiments. But now in this quiet retreat beside the Grand River, with returning health he began to give more thought to the problem of the Multiple Telegraph, as it was commonly called; and it occurred to him that with his knowledge of music, he might make use of what is known as "sympathetic vibration," the fact that when a given note is sung into the piano close to the keyboard with the pedal depressed, the corresponding key of the piano, tuned to the same pitch, will answer. And because his solution of the problem involved the use of musical notes he spoke of the device which he had in mind as "the harmonic telegraph." Aleck spent long hours at the piano, putting the theory of sympathetic vibration to the test; but these experiments were not quite so simple as they looked, and progress was very, very slow.

CHAPTER TWELVE

The War Dance

WHEN MR. HENDERSON was at Tutela Heights some weeks later, he happened to mention the Six Nation Indians.

"Who are they? What are they like?" asked Aleck.

It had never occurred to him that any of the North American Indians about whom he had read so much in history, might be living right here close beside him; and Mr. Henderson had never thought of mentioning them.

"They live over there," he hastened to explain, "on the other side of the valley on the Indian Reserve. You can almost see the old Mohawk Church from here. Any Saturday when you're in town you'll find plenty of them hanging 'round the street corners."

"Who are they? That's quite a long story. They're the Iroquois—what are left of them. For hundreds of years their ancestors lived in the country to the south of Lake Ontario. But when the British were defeated in the American War of Independence, the Six Nations, who were their allies, lost all their lands. To make up to them for their losses, the British set apart a tract of land, on each side of the Grand River, for them, and under their leader, the great chieftain Joseph Brant, they settled all along the river here for a distance of twenty or thirty miles. The land where

Brantford stands was part of the Six Nations Reserve, but they sold it to the white man, and here the city was built."

"Why do they call them the Six Nations?" Aleck inquired.

"Because there are six tribes," replied Mr. Henderson. "The Mohawks (the strongest tribe), the Onondagas, the Cayugas, the Oneidas, the Senecas, and the Tuscaroras."

"What does their language sound like?" asked Aleck. "That's what I'm interested in."

"I can't tell you that," replied Mr. Henderson. "It's just Indian to me. What about your Visible Speech?"

"I was thinking of that," said Aleck, "and I'd like to write it down. But I'm afraid I can't do that for a while yet. Would they mind if I went over there sometime when I'm a little stronger? I'd like to talk to them."

"No," said Mr. Henderson. "They'll be glad to see you. I'll introduce you to Mr. Gilkison, the Government Agent, who is their guardian. He lives near you. He'll take you over, or you can drive yourself. It's five miles to the Old Church and six or seven miles further to their Council House."

Some weeks later when Aleck and Mr. Gilkison happened to meet on the Market Square and stopped to talk, Mr. Gilkison suddenly broke off in the middle of a sentence and pointed to an Indian who was crossing the street.

"There he is!" he exclaimed. "That's the man! I was hoping we'd meet him."

"Who is he?" said Aleck.

"Everyone knows him," continued Mr. Gilkison, "Chief Smoke Johnson, the best of all the Indians. He'll help you, if anyone can. His wife is an English woman, from Bristol. They live down the river, ten or twelve miles from here in the big white house called 'Chiefswood.' You must go down there some time. He'll tell you all you want to know, in English or in Mohawk, whichever you please. I wish Joseph Brant were alive, he could talk with kings and princes, the proudest in the land."

Chief Smoke Johnson was interested in the dark eyed youth, and one day when Aleck and his father journeyed down the long sandy road to Chiefswood, and sat down to a dinner of wild fowl, they talked a long time about the Indians, the Neutrals, the old Indian town of Kandoucho, where the city of Brantford now stands, the Grand River, and a hundred other things.

The chief's wife was a woman of education and refinement. She spoke of her girlhood in Bristol and her first years in Canada, and all the time her little seven-year-old daughter, Pauline, clung shyly to her mother's dress or hid behind her chair, finger in mouth, too bashful to speak a single word to the two bearded strangers. But if Aleck Bell had only been able to look into the future and follow the fortunes of this gifted child of the Reserve, he might, in another twenty years to come, have heard the voice of Pauline and seen her in her Indian costume as she chanted "The Song My Paddle Sings," or recited the death song of the Iroquois chief in her poem "As Red Men Die."

As the long months came and went, Aleck became more intimate with the Mohawks and visited them in their village, or talked to them on a market day, on a busy street corner of Brantford. He was always "a little queer," said the gossips, and here was further proof if any were required. The Indians were pleased with the notice that he took of them, and were interested in his attempts to write down their language in Visible Speech. With every visit that he paid to them they were the more flattered, and they showed their pleasure in later years by inviting him to become a Chief of the Six Nations Confederacy.

The ancient ceremony took place in the Council House at

Oshweken on the Reserve, over which the Union Jack was flying to show that the Council was in session. The Chiefs were seated behind a railed-off space, with the Onondagas, "the Fire Keepers," in the centre. While the new Chief was being introduced, the Chiefs stood. Then they invited him to sit in Council with them, while they debated some matter of common concern to all the tribes. The debate was opened by a Mohawk Chief and when an agreement was reached, the Onondagas were required to give their sanction. Then the new Chief received his Indian name, and was taught the war dance. After this the pipe of peace was passed, and the Council was ended, amid piercing whistles and war-whoops. Then the new Chief, Alexander Graham Bell, decked out in all his finery, and with the wild shouts of the war-dance ringing in his ears, made his way back to "Melville House" and turned once more to another round of patient experiments with the tuning forks and the Multiple Telegraph. The war-dance served him well in after years as an outlet in moments of wild enthusiasm. When some signal success had crowned his efforts, a wild war-whoop and a whirling dance was a relief to his overwrought and pent-up feelings.

Part 3.

"TO BE, OR NOT TO BE"

CHAPTER THIRTEEN

The Plunge

IN THE MEANTIME Aleck's health continued to improve and Professor Bell saw no reason now why he himself should not go to Boston again to deliver another series of lectures; and in October he returned there to fill this engagement. When this course of lectures was concluded he was asked to instruct the teachers of the deaf in Visible Speech. Unfortunately another long standing engagement prevented him from doing so. "But," he suggested to the committee, "if you care to send my son Aleck an invitation, he is quite competent to take my place."

But in spite of Professor Bell's recommendation, the committee were at first a little dubious. The young man was only twenty-three. It was natural that his father should put in a good word for him, but a man so young as he, they urged, could scarcely have mastered the scientific principles involved in this difficult subject! They did not want to run the risk of being disappointed; but after much discussion they sent him an invitation and he accepted it. He was glad to have some respite from life on the farmstead; and even more important still, the Board of Education in Boston had voted the sum of five hundred dollars for payment of the lecturer!

Aleck did not go to Boston wholly as a stranger. His father had spent a happy month there until the cruel blow fell which bereft him of his eldest son and hurried him back to London. He had made many friends in Boston even in the short time that he had been there, and he had written long letters home, telling the boys and their mother about the people and about his lectures in this far-famed "City of Letters."

In his school days Aleck had read about the Boston Tea Party and about the ships that sailed out of the Boston harbour with its ring of hills. He knew of Faneuil Hall, the cradle of liberty, and the State House and the Old North Church, from which Paul Revere had set out on his midnight ride, and Boston Common, set apart two centuries before as a soldiers' parade ground and as pasture land for sheep and cattle. When Aleck emerged from the railway station it was with a feeling of exhilaration.

To his audience of teachers he must have seemed an unusual type of lecturer. He was a mere lad, a stripling, tall, and as yet, slight of build, face pale, full lips, and big Bell nose, high sloping forehead, side whiskers, after the fashion of the day, and jet-black hair, through which he kept running his fingers till it stood on end. His movements were quick and his gestures animated; and he spoke in distinct crisp tones, with no trace of accent. His father and grandfather had seen to that. It was not for nothing that Grandpapa Bell had taught him Hamlet's advice to the players.

"Speak the speech, I pray you, as I pronounced it to you, trippingly on the tongue; but if you mouth it, as many of your players do, I had as lief the town-crier spoke my lines."

The course of lectures covered a period of six weeks or more, and they were so successful that during the next few weeks he received invitations to lecture in half a dozen other schools for the deaf. But in the course of a few months he was back home in Brantford once more, continuing his experiments and working from time to time on the "farm." The winter months passed slowly. There was little need to go into town except for the purchase of supplies, and apart from his experiments there was little for Aleck to do. He had a long rest, and before the winter was over, he had, to all appearances, quite recovered from his illness. But he was restless and with the first mild days of Spring, he longed to be out of doors. "I can't turn Indian," he said, "but I can't sit here for ever, like Micawber, waiting for something to turn up. I must find some employment, but what can I do? There's nothing for me here."

"I must be a teacher of speech," he continued. "That was settled for me before I was born, but where? And when? And how? I can't stay here all summer doing nothing, waiting, waiting, waiting. It isn't fair to you," he said to his father and mother, "that I should live under your roof so long, and idle away time." Then, after a moment's pause, "Do you know, I've half a mind to go back to London, now that I'm quite well again. You are well-known there. When we left for America, I was making friends, and I could easily earn a living."

"London?" exclaimed Professor Bell, raising his eyebrows. "You're surely not in earnest. It would be madness to think of going back there. You had a narrow escape, and we must not risk another illness. In any case we should have to start all over again, and we can't afford that. When we bought this place we planned to stay here for two years at least. Besides, I like this country. It's a better place to live in than Old London with its dust and its fogs, but of course it is not half so interesting or exciting. But let us change the subject. What about this Multiple Telegraph you are working on?"

"'Harmonic,' I call it," said Aleck. "I can't be sure of it. It would make us rich; but I don't know enough about it. I have scarcely made a beginning. I'm hoping that something may come of it, hoping, that's the best I can say. But these things take a long time. For one man who succeeds, there are a hundred who fail. I'd like to have a few weeks more to work on it."

But in six weeks' time the problem of the Multiple Telegraph was no nearer to solution.

"I don't know enough about electricity," said Aleck, "but I can carry on my experiments while I am teaching. If I cannot go back to London, what about my going back to Boston? I have an idea that if I advertised for pupils I could be reasonably sure of making a living. I'd like to open a school there for children who stammer or who cannot speak plainly, the deaf and the dumb, who can neither hear nor speak. There are plenty of them in Boston. But I'm afraid I might fail."

His father looked up sharply. "'But screw your courage to the sticking place,'" he replied, "'and we'll not fail!'" There was a note of grim decision in his voice. It was not the first time that Shakespeare had come to the rescue of the Bells!

Professor Bell wrote to friends in Boston for assistance and advice; and towards the end of September, 1871, Aleck left the quiet hill-top at Tutelo Heights and set out for Boston to become Mr. Bell, Professor of Correct Speech—"Vocal Physiology" they called it. In so far as Aleck's future was concerned, it was an all-important decision. The next three or four weeks were busy ones for him. To his great delight there was a large enrolment of pupils, and the new school bade fair to be a crowning success.

One of his first pupils was a little lad named Georgie Sanders, whose father was a well-to-do leather merchant of Haverhill, Massachusetts. Georgie had been born deaf, and as a result had never learned to speak. His father had put him under the care of a well-known teacher, and this lady wished to have Professor

Bell take charge of his education. He was a wistful and lovable boy, and the two soon became greatly attached to each other. Georgie, with his nurse, went to live in the boarding-house in Boston in which Bell roomed; and during the winter months he spent many long lonely hours waiting for his friend and instructor to return from his school to give him his daily lessons, and have some fun! The tall man with the black eyes and the tousled black hair was only a big boy after all!

"Little Georgie is progressing splendidly," wrote Aleck to his mother. "He is a loving and lovable little fellow. I wish you could see him. I expect great results with him if I can have him with me for two or three years."

During these months Aleck was under great nervous strain, teaching his classes and private pupils during the day, and spending long nights in experimenting with apparatus that might help him solve the problem of the Multiple Telegraph, and might bring in some ready money. All his life long he was unbusinesslike with regard to money. Now he found his classes dwindling because he was unable to pay proper attention to them, and if it had not been for the payments that Mr. Sanders made to him as Georgie's tutor he would have fared very badly.

During this winter he made the acquaintance of another man who later proved to be a friend in need. He was a lawyer, named Gardiner Greene Hubbard. His daughter, Mabel Hubbard, had become deaf at the age of four, as a result of an illness. She had been taught by an American teacher, under her mother's supervision, to speak, and to read the lips. This was a very new thing at the time and she was one of the first two children in the United States to learn lipreading. Then her parents had sent her to Germany to see if her speech could be improved. The teachers in Germany were amazed at her proficiency. And there she learned to speak and read German and to read the lips of Germans. Because of her affliction, her father had become interested in the education of the deaf, and particularly in the work that young Mr. Bell was carrying on in this school of his. Now that his daughter had returned from Germany, he had arranged that she should receive further instruction under the direction of Mr. Bell, who had taught his pupils the Visible Speech alphabet with such wonderful results. Mr. Hubbard was ready to do anything and spend any amount of money for his daughter.

In the meantime Aleck was finding the year's work too heavy for him, and his health was beginning to suffer. His mother's weekly letters were full of inquiries about the cause of his headaches. Did he exercise enough? What about the height of his pillow? On hot days did he ever think of putting a leaf of rhubarb or of cabbage in his hat? In cold weather did he wear the flannel bed-gowns she had made for him? Was it possible that he played the piano too much in the evenings before going to bed? And then, as his letters revealed his preoccupation with his work on the Multiple Telegraph, she wrote, "We will talk of your invention, my dear boy, when you are at home. I trust it may answer the purpose intended, but I want you not to think any more about it just now. 'A bow always bent,' you know, 'will break.'" And soon she was urging, "Don't allow yourself to be made nervous; these ups and downs come continually in life, and if your mind is collected, you will the more readily see how to act for the best." She could hardly realize that he had undertaken too much and was burning the candle at both ends. He thought that the winter would never end, and he was glad when summer at length came and he was able to leave behind him the hateful boarding-house on West Newton Street, and turn his face once more to the little town on the Grand River, in Ontario. There he would relax and see all the improvements that Papa had made in the home stead—energetic Papa, travelling about, giving readings in Toronto and Kingston and in many smaller towns, yet finding time to make a croquet ground, to build a new porch on which the vines would now be in full leaf, to build a fence to keep the fowls from the garden—and even to renovate the buggy.

And there would be his mother's flowers to admire, some of them grown from seed and roots which he had sent to her. He would wander out of doors and see the wild flowers, specimens of which his mother had put in her letters. "Our surroundings are very lovely just now," she wrote, "but we feel the charm to be incomplete, when you are away." And so he went home, his only regret being that he had to leave little Georgie behind.

CHAPTER FOURTEEN

The Magic Glove

THE ARRANGEMENT BY which Georgie Sanders lived at Mr. Bell's boarding-house in Boston and was taught by him in his leisure moments, was not very satisfactory. Georgie was left alone all day long with no one but his nurse and the landlady for companions; and the nurse and landlady did not always look after him as they should. Why should they trouble? Mr. Bell was away most of the day, and Georgie couldn't tell tales. This sort of life was not good for the child, who could not hear a sound and who was just learning, slowly and painfully, to speak a few words.

"Something will have to be done," said Mr. Sanders to Georgie's grandmother. "I want Mr. Bell to teach him. He is doing wonders with the boy, and they think the world of each other. If he stays with Mr. Bell long enough he'll be able to talk to you. But a five year-old boy shouldn't be left to himself so much of the time."

Mrs. Sanders, Georgie's grandmother, lived in Salem in a roomy big house, and she was sometimes very lonely.

"Could Georgie live with me?" she asked, rather hesitatingly, "and Mr. Bell must come too, of course. Georgie couldn't get along without him. He'll have to take the train into Boston every morning, but it's not so far, only fourteen miles. I'll have to give him his breakfast, and his supper too, perhaps, and we'll have to arrange about Georgie's lessons. Do you think we could do it?" In the end Mr. Sanders decided to undertake it, and that was just what Georgie's grandmama was hoping for. In the old days when superstition poisoned even good men's minds, she might have been burned at the stake, or drowned in the horsepond, for taking a "magician" into her house. Salem was a city of ghosts, but the good lady, kindly soul that she was, was in no danger now.

When Aleck returned from his visit home, (October 1873), and heard of the new arrangement, he was delighted. He had been appointed Professor of Vocal Physiology in the School of Oratory at the University. This meant that he would have to go into Boston every day, an hour's journey by train, to teach his classes, and it wasn't pleasant to have to catch an early train on a dark blizzardy morning—and New England winters are bitterly cold; but in Boston he could take a room near the University, and he would be free of the boarding-house, he hoped, for ever. It was worth trying.

But there was one serious drawback to the arrangement. Where would he carry on his experiments? He would have to have a room either in Mrs. Sanders' home, or nearby, in which to work. After returning from Boston and spending an hour with Georgie, he must be free for the rest of the night to work on his Multiple Telegraph. Mrs. Sanders, generous and grateful to Mr. Bell for the help he was giving to Georgie, gave up the whole basement to him. Mr. Bell was "such a nice young man, and so considerate."

It was no small thing for an elderly lady to put up with the noises of a workshop for most of the night, even for the sake of Georgie. But the generosity of Mr. Sanders, Georgie's father, went still further. Not only did he pay Aleck for teaching Georgie, but he undertook to pay for the supplies that he used in making his experiments. Aleck himself had no more idea of the value of money than a child, and not so much as most children.

But the one thing that Mr. Sanders could never pay for was Aleck's affection for the little lad, who lived in a world all his own into which there penetrated not a sound, not the song of a bird, nor the joyous bark of a dog, nor the shout of a companion, nor the reassuring voice of the kindly teacher, the most wonderful grown up boy, who came home to him every evening on a noiseless train moved by a great puffing engine that didn't make a sound. Little George Sanders heard not a syllable of what the engine said. The train wasn't even a "choo-choo" to him, but just a big black monster that frightened little boys and girls. He had been deaf from birth, and until Mr. Bell came, he could make only unintelligible sounds. But Bell with infinite patience taught him the shapes of words, so that one by one he was able to pronounce them and read them.

He was a quick, intelligent lad, and no better teacher could have been found in all the world than young Mr. Graham Bell. It was an untold delight to the boy when he was able to recognize and

pronounce words; but there was a still greater source of pleasure in store for him. When he had learned the letters and simple sounds of his alphabet, Aleck one day brought him a leather glove for his left hand, on which letters and simple words had been printed, and the little boy spelled out "Thank you, Mr. Bell."

Georgie used his forefinger, pointing to the letters on the glove to spell out the words that he wished to say, and Aleck talked back to him in the same way. It was wonderful how fast they spoke and how well they understood. On his journey home each day Aleck thought of things that had taken place during that day, and tried to call to mind little stories that he had read in the daily papers, that he could tell to Georgie, children's stories to which the magic glove was the key and that didn't have any noise in them. For the boy, this study hour was the happiest time of the day. For Aleck himself it was an hour of utter relaxation. It took him out of himself and he gave no thought to Multiple Telegraph or Visible Speech; and it was not until the little boy had gone happily to bed that he turned once again to the magnets and reeds and wires and membranes out of which he hoped someday to wring hidden secrets, and perhaps, too, a mountain of gold.

If anyone had asked Mrs. Sanders what the "Professor" was working at during these long winter nights, she would not have been able to tell. She didn't know what his experiments were about. She knew only one thing, that he was teaching Georgie to speak and that the child was perfectly happy when Aleck was there; and she knew that his experiments had something to do with the piano, and with wires, for he seemed to have them strung all over the place from one end of the basement to the other, and up to the attic floor and even across to the neighbour's upstairs window. She knew nothing of the Multiple Telegraph on which he was working and out of which he hoped to make a fortune, nor of any of the other plans which he had in mind. She only knew that he was "such a nice young man, and so considerate," and that Georgie was perfectly content when Mr. Bell was telling a fairy tale to him with the magic glove.

CHAPTER FIFTEEN

"A Real Ear?"

WHILE ALECK WAS carrying on his experiments during the winter of 1874, he made a new acquaintance who was to play an all-important part in his life. In those early days in Boston, young inventors who wished to have some machine or some part of it, some "contraption" on which they were working, made up, usually took it to the machine shop of Charles Williams, who employed a number of young men on work of that sort. When Bell was working on his Multiple Telegraph he one day took a piece of work back to have an alteration made; but instead of going to the office desk with it, as he was supposed to do, he dashed right into the machine shop, after his impulsive fashion, and went over to the work-bench of the young man who had made it. He was a mere lad, only twenty years of age, some years younger than Aleck, somewhat shy, but a rapid and accurate workman, and a modest and likeable youth. His name was Thomas Watson.

Aleck explained the invention on which he was working and as Watson made the required parts for him, from time to time they tried them out together. The two enthusiastic youths became firm friends even though they differed totally as to background. Bell was a cultured young man from an exceptional home, while Watson was without any advantage of education or family. "No finer influence than Graham Bell ever came into my life," Watson wrote towards the end of his life. And he tells how Bell's punctilious courtesy delighted him, how he imitated Bell's table manners, and how Bell introduced him to the writings of great scientists such as Tyndall and Huxley. But what fascinated Watson most was Bell's mastery of expressive speech and of piano playing. Aleck gave Watson some of his father's books on Elocution—and corrected Watson's errors of speech.

Watson soon discovered that Aleck's mind teemed with other new ideas—besides that of the Harmonic Telegraph. "A dozen young and energetic workmen would have been required to mechanize all his buzzing ideas," said Watson. One of the inventions that he was working on was a device which might prove to be a help not only to children like Georgie Sanders, but even to grown-ups who were deaf and dumb. But yet, as it turned out in the end, it did have a great deal to do with solving one of the problems of the telephone.

In listening to the voices of people who have been born deaf and have been taught to speak, he had observed that their tones are monotonous and disagreeable. Never having heard anyone say a single word, they cannot imitate the sound of the human voice, and they have no idea that they can play upon it as if it were an instrument. George Sanders, like all other deaf children, spoke in a harsh unnatural tone; and as Bell listened to his attempts to speak, he was moved to pity. "If it were only possible," he said to himself, "to teach these deaf children how to modulate the voice to show pity, anger, fear, and a thousand other shades of feeling, the world would be a much happier place for them and for us to live in." He set himself to work eagerly to try to find some way of making this idea of modulation clear to them. "Since these children cannot hear," he argued, "I must make use of something that they can see."

He was aware of two devices, both recent inventions, which were intended to show in visible form the changes in the quality of sounds. The first of these devices was known as the "manometric capsule"—a clumsy name until you get used to it—in which a gas flame rises and falls in flickering waves according to the intensity of the tones of voice, and in so doing, forms a wavering band of light. But Bell found this capsule of little use to him because in those days it was impossible to photograph these flickering light waves so as to show the changes that took place.

But the other device, it seemed to Bell, might be very useful. It was called the "phonautograph"—another clumsy name—and it consisted of a mouthpiece, a stretched membrane to which the end of a stiff bristle was glued, and smoked glass on which the free end of the bristle made a pattern when the membrane vibrated to the changes in the intensity of the voice. Bell made repeated experiments with the phonautograph, and in so doing he was struck with the resemblance between the vibration of the membrane and bristle and the action of the eardrum and the bones of the human ear. It seemed to him that if he could make his receiving apparatus (the membrane with lever and bristle) resemble the ear more closely he might secure still better tracings on the smoked glass. Bell had a friend, a doctor named Blake, who was interested in these experiments with the phonautograph, and the two of them talked the matter over together. Blake listened to what he had to say, then asked almost casually, "Why don't you use a real ear?"

"A real ear?" Bell gasped. "Where could I get one?"

"Don't trouble about that," replied Blake. "I'll get one at the Medical School from a corpse."

Doctor Blake was as good as his word. He procured a human ear with the inner bones attached, and with this unusual equipment Aleck continued his experiments. He had the idea that his smoked-glass pattern would show how the voice rises and falls in speaking, and that a deaf child might learn to modulate the voice so as to make similar tracings on the smoked glass. But he was, as usual, too hopeful, and as far as deaf children were concerned, the experiments with the phonautograph came to nothing.

But as he studied the "sound recorder" itself, with the ear of the dead man as part of it, he was struck with the fact that a membrane so slight and thin as the eardrum, as thin as tissue paper, could make the heavier bones of the ear vibrate. "If," he reasoned with himself, "the eardrum, which is so small and delicate, can set in motion the bones of the ear, which are so much heavier, will not a thin disc of iron act upon a much heavier lever of metal and set it vibrating?"

This was just the hint that Aleck needed in order to complete his idea of the telephone: a mouthpiece, a diaphragm (membrane, eardrum, iron discs), an electrified arm of steel (lever, wire, earbones), a receiving diaphragm, to transmit the vibrations; and so there came into being the idea of the modified phonautograph—the forerunner of the telephone.

When Aleck set out on his journey home to Brantford a short time later, for the rest he so much needed, the gruesome relic, the human ear, was part of the apparatus which he carried with him. If, that summer, the neighbours had seen and heard him shouting all day long into a dead man's ear, and getting no answer at all, they would have been doubly certain that he was "just a little queer."

CHAPTER SIXTEEN

The Dreaming Place

At the top of the steep bank above the river, a hundred feet from the Bell home in Brantford, stood a little grove of birches. Between two of the trees a hammock had been slung, and this quiet spot was what Aleck called his "dreaming place." Up in his bedroom he shouted into the membrane of the human ear and made tracing after tracing on his smoked plates. Then, resting in his hammock, he turned over in his mind what he had learned from the tracing, and tried to reason out the problem of transmitting human speech over a wire by the use of electricity—electric speech.

To a man like Bell who had already carried on so many experiments in sound, the thing looked simple enough. A wire with an electric current should, it seemed to him, pick up the vibrations of the voice from the transmitter, carry them along, and set sound waves in motion at the receiving end of the wire. A dead wire would, of course, carry nothing. The electric current was the thing! But Aleck had already learned from his experiments that an ordinary current of electricity would not carry sound, least of all sounds so complex as the human voice. And if not, what sort of current could do it? Point by point and piece by piece he reasoned it out. In order that an electric current sent over a wire could carry sound waves, it must vary as the air varies when a person shouts or whispers or laughs or strikes a note on the piano. It must be possible to generate a current like that! But how? Until that question was answered, there could be no such thing as electric speech, except in the brain of "Crazy Bell," as some folks called him. Transmitter, receiver, and to "go-between," the current! He tried to put his idea into simple language; but even then it sounded to the ordinary man like a scientific jumble. "What I want to find," said Bell, "is a mechanism which will make a current of electricity vary in its intensity—as the air varies in density when a sound is passing through it. With such a mechanism I can telegraph any sound, even the sound of speech."

Yes, that covered the case. There was no doubt of it in his mind. Here in his "dreaming place" he had put his finger on the solution

to his problem. He must find a current, or create it, that would vary in the same way as the air that carries sound waves varies. The current must *undulate* (that was a word to look for in James Murray's new dictionary. Latin, *unda*, a wave).

Bell was elated. He knew that he had taken a great step forward, but at times he felt some misgivings and was greatly cast down. This undulating current— did such a thing exist? And how could it be created? This was the key to the problem of electric speech but he didn't know how to unlock the door. This undulating current was an elusive thing. How would he recognize it when, or if, he did come upon it? Elisha Gray, his rival, or some unknown inventor, might come upon it first! That would be a body blow! If this theory of his proved to be sound, and if it could be carried out into actual practice, it would mean that he had already invented the telephone, or at least worked out the formula for it. It was here, as Bell himself said, here, under the grove of birch trees at Tutelo Heights, that the telephone was "conceived," if not born. And later in life, with the help of his father's diary, he went so far as to fix the date when the idea first took definite shape in his mind as August 10th in the year 1874, the year of years, and the day of days!

It seemed to Bell as he returned to his problem, that there could be no slip in his reasoning, but he went over it time and again, step by step, to make sure that he had forgotten nothing. He was secretly excited! Each morning when he woke, he was filled with eager hopes, and with vague apprehensions; and with each experiment that he undertook he felt that he might, at any moment, stumble upon the secret. He went back to his experiments on the Multiple Telegraph with renewed courage.

CHAPTER SEVENTEEN

"A Little Mad"

WHEN, IN THE AUTUMN of this year, 1874, Aleck returned to Salem from his long vacation in Brantford, he found that Georgie and his grandmother had not yet come back from Haverhill. Aleck, eager to see the little fellow, went on to Georgie's home there. In a letter to his mother he described their meeting. "You would have been amused to have witnessed his reception of me. He is generally very undemonstrative in his welcome to friends; so I felt quite flattered by his delight. His father and mother were out driving at the time but when they came home, he made me hide behind a door that we might surprise them." And he continued, "Mrs. Sanders is hard at work as usual making me feel at home. She is the very essence of kindness."

Aleck's mind was full of his new idea of Electric Speech. He could think of little else. But he was hard pressed for money to meet his ordinary expenses; and he was forced to give some thought to his classes, and to his lectures for the college year which was just about to begin. He could not very well talk about his personal affairs with his casual acquaintances, and he had no very intimate friends with whom he could discuss his work and his plans.

One day in the late Autumn, Aleck received an invitation to have dinner with the Hubbard family on the Sunday following.

This was something of an event for him, for he had few friends to whose homes he might be invited. The Hubbards lived in Cambridge, a residential suburb of Boston. The Hubbard home was a spacious mansion, placed in a park-like setting of five or six acres, and furnished comfortably, if not luxuriously, in the fashion of that day. Aleck accepted the invitation gladly, the more so as he secretly hoped that he might have a half-hour's chat with the second daughter of the house, Mabel, about whose speech her father had consulted him. For the interest this lovely and talented young girl had aroused in him on their first meeting was becoming much more than a professional interest in her speech; he was rapidly falling in love.

Mr. Hubbard was a shrewd, hard-headed, businessman, something of a philanthropist, and a man with a wide range of interests. In the two years that he had known Mr. Hubbard, Aleck had not had the courage to talk over his plans with him for fear that the matter-of-fact "promoter" would sweep them away as impracticable. But, as he found out in due time, Mr. Hubbard had a romantic and somewhat adventurous side to his character, which served Aleck in good stead later on.

In the afternoon Aleck sat down to the piano, even though he knew that Mabel Hubbard could not hear and enjoy the music; but in the midst of the sonata he suddenly stopped and, turning about abruptly, said to Mr. Hubbard:

"Do you know that if I sing a note into the piano, at the same time depressing the pedal, the corresponding piano string will answer me?" and Aleck suited the action to the word.

"Well," said Mr. Hubbard, "what does that matter—"

"It matters a great deal," said Aleck. "It means that when a note is sounded at one end of a wire, the corresponding key at the other end of the wire will answer; and if I strike half a dozen

In the afternoon he sat down to the piano.

keys at A, the same wire will carry these half dozen sounds to B, that is, if the keys at A and B are tuned to the same pitch. The receiving key vibrates in sympathy with the sending key (or voice). This is something that every schoolboy knows; but it has not been applied to the telegraph. If it can be applied, it means that a single telegraph wire may carry half a dozen messages, or a hundred if need be, at the same time; and the problem of the Multiple Telegraph is solved."

Mr. Hubbard, who had been listening to the sonata with half-closed eyes, became suddenly interested. "Keep that in mind," he said to Bell. "It is worth working at. You may make something out of it—you never can tell—and I'll pay the cost of your experiments."

"Thank you, sir," said Aleck, the blood flooding into his pale face, "but Mr. Sanders has already promised to do that for me. I've been working on the Multiple Telegraph for some years, but I have in mind a better idea. I think that the human voice can be transmitted by electricity over a wire. I wish I had the time to experiment with it.

"Now you're talking nonsense," said Mr. Hubbard. "It is absurd, utterly absurd. But the Multiple Telegraph is another thing. I'll speak to Mr. Sanders about sharing the expenses. In the meantime put all these other idle fancies of yours out of your mind, and work at this."

That evening at Mr. Hubbard's, Aleck seemed a little absent-minded. A good many other thoughts were running in his head; but on one thing he had firmly made up his mind, he would not give up the idea of Electric Speech no matter what Mr. Hubbard might say.

The one man to whom he felt he could talk freely was Georgie's father, who had shown faith enough in him to finance his experiments; and he told Mr. Sanders about his dream of Electric Speech. Sanders was, naturally, not very enthusiastic. He had already advanced a great deal of money, with nothing as yet to show for it; and he thought that Aleck should begin to think of how to pay it back, instead of spending more. A bird in the hand—the Multiple Telegraph—was worth half a dozen in the bush. Bell's enthusiasm was dampened, and he decided not to say anything more about it to Mr. Hubbard. Electric Speech, if it ever did come to pass, would be a tremendous world-shaking thing—of that he was certain—but to a hard-headed practical man all such schemes were but toys and playthings for an idle hour; and he, Alexander Graham Bell, must have seemed to them to be a mere dreamer. Either these men were blind or he himself was a little mad. He would, he promised himself, give more time to the Multiple Telegraph—he owed that to Mr. Sanders—but once again he made the resolve not to give up his cherished dream of Electric Speech.

It was clear to him that he need not look to either Sanders or Hubbard to encourage him in the pursuit of this will-o-the-wisp, but what about Watson? He was really scarcely more than a boy, a practical fellow, yet not wholly matter-of-fact either, a youth to whom other dreamers came with their harebrained fancies. He had seen too many of them fail not to be doubtful of any new inventions that hadn't been thoroughly tried out. But Mr. Bell was different from the rest of them, not a slow plodding matter-of-fact machine-made type of man, but over-impulsive and much too sanguine for his own good, too much up in the clouds! Watson was shy with him at first. The practical man doesn't understand the dreamer any more than the dreamer understands him. Besides he (Watson) had kites of his own to fly, dreams, most of them, too, and he wanted to talk to Aleck about *them*. But he was tongue-tied, and left it to Aleck to do most of the talking.

The two of them sometimes stayed in the shop after hours to finish up a piece of work; and there were times when they stayed in Boston overnight to try out a piece of mechanism when the factory was quiet and when they had the place to themselves. One night when they had stayed later than usual and had stopped for a few minutes to rest, Bell said to Watson, "I'd like to tell you about a scheme I have in mind, a new kind of electrical machine that will carry sounds of all kinds over a wire, not just a 'make-and-break' kind of signal, but real words, your voice and mine. Think of it, Watson! It means that we could talk to our friends across the road or in another town or half round the world—and it's no dream either. I have the idea, but I can't find time to work it out just now; when I get this Multiple Telegraph going I'll have enough money, and time enough, for all sorts of things."

Watson was startled by the very boldness of the scheme. Talk over a wire! The wheels ran faster in his head! If that dream ever came true, the whole world would be different. "If!" But who could say? It seemed incredible, but this Bell was an ingenious fellow, and somehow he stirred things up inside of you and set you thinking. A little mad—but interesting and agreeable!

CHAPTER EIGHTEEN

"Get It!"

In the meantime the Multiple Telegraph was getting on. Bell had applied for a patent, and one day the following February he received a message from his patent attorney that he wished to see him to talk over the specifications with him. That meant a journey to Washington. Bell had never been there; but he was taking steps to become an American citizen, and Washington meant something more to him now, a great sprawling infant of a city, still in swaddling clothes, not quite a century old, for the growth of cities is not measured by years but by epochs.

But America was on the march and things were being done, and would be done, in a big way. The shadow of Lincoln was over this new world, his kindliness, his foresight, his broad humanity, a world figure. It was a different kind of city from Edinburgh and London and Boston, for somehow the spaciousness and freedom of the new world was here. The impulsive, impetuous youth could feel the thing in his blood, and he knew that he, too, would have a part in it. Hence forward there were two worlds for him—the quiet little town on the "Rapid River," the Mohawks, and the hammock under the birch trees, and on the other hand, the Capital City of the new world and all it stood for, to which Boston was a sort of way-station.

This summons to Washington could not be ignored. He was

desperately hard up; but he must find the money somehow, somewhere. When Mr. Hubbard, who had lodgings the year round in Washington, asked Aleck to make use of them he little knew how very welcome his offer was.

The appointment with the patent attorney was soon disposed of, but Bell had another and more important interview on hand. The greatest authority on physics in America, or perhaps in the world, was Joseph Henry, the Secretary of the Smithsonian Institution. He was the one man in Washington who could give Aleck advice on the experiments he had undertaken. Could he, Aleck Bell, make bold to call and ask for an interview with so distinguished a scholar? He hesitated, and it was with some trepidation that he made his way to the Smithsonian Institution.

Joseph Henry had devoted a long lifetime to the study of physical science; but he listened patiently while the young man described his experiments; and he was particularly interested in those in which Bell seemed to have made original discoveries. The long afternoon was passed in Bell's demonstrations with his own apparatus which he had brought with him. Then at the end, after question and answer concerning his Multiple Telegraph, Aleck ventured to speak of his dream of Electric Speech. Henry was all eagerness. There was half a century between their ages, half a century of careful investigation and experiment on the part of the older man; and for the other, half a century of achievement still to come.

When Joseph Henry had heard the eager young man through to the end, he said to him, "You have there the germ of a great invention. Work at it." Then Bell, with the humility that belongs to the great, expressed the fear that he had not enough knowledge of electricity to carry it through. "Get it," said Henry. In his record of that day's activities, Bell wrote the words "Get It!" in capitals, and writing home to his father and mother he said, "I cannot tell you how much these two words have encouraged me. Such a chimerical idea as telegraphing vocal sounds would to most minds seem scarcely feasible enough to spend time working over. I believe, however, that it is feasible and that I have got the clue to the solution of the problem."

"Get it! Get it! Get it!" The very rails and engine with its heart of iron sang in his ears. "Get it, get it—get it!"

CHAPTER NINETEEN

Stress and Strain

AFTER HIS RETURN to Boston, Aleck wrote to his parents to tell them about his visit to Washington, and in the course of his letter he mentioned incidentally that Mr. Orton, the head of the Western Union Telegraph Company, had called to see him and had looked over his Multiple Telegraph apparatus; and he spoke of having met Orton a second time a couple of days later. He was greatly elated; and why not? The Western Union was the largest and wealthiest corporation of its kind in America; and if Orton should report favourably on Aleck's invention, the young man's fortune was made.

Mr. Orton was very cordial and invited him to take his apparatus to New York in order to consult the Company's electrical experts regarding it. Aleck had been trying for months past to interest the Western Union in his Multiple Telegraph; but they had given him no encouragement. And now the tables were turned, and they had come to him and had invited him to demonstrate his machine to them!

"Hurrah!" he cried aloud to Watson. "It's as good as sold! The Western Union want to see it. I am to demonstrate it in their offices in New York."

Aleck, as usual, allowed his enthusiasm to run away with him. But Watson threw a dash of cold water on it.

"Don't be too sure!" he cautioned. "I've seen this sort of thing happen too often before. You can't be sure of anything until the last 'i' is dotted and the last ' t' is crossed."

Aleck went to New York the following week, and gave the demonstration. The experiments seemed to him to be satisfactory, but the Company's experts were not very greatly impressed. The invention, they agreed, was sound in theory, but they were of the opinion that the harmonic machine would get out of tune too easily and that if it were not constantly adjusted, it would give no end of trouble.

Aleck was disappointed. The Company's expert had put his finger on the weak spot in the invention, and Aleck himself knew only too well that the harmonic telegraph was not likely ever to be a complete success. He continued his experiments, but both he and Watson worked half-heartedly. The truth was that they were now much more interested in Electric Speech than in the Multiple Telegraph.

Sanders and Hubbard, strangely enough, still had faith in the Multiple Telegraph, and they went so far as to have an agreement drawn up by which they undertook to meet all expenses of further experiments and share all the profits if it proved successful. But the results of these experiments were not encouraging. Nothing went right. The Multiple Telegraph continued to give them trouble, and Aleck's experiments with the phonautograph and with the dead man's ear seemed to have accomplished little or nothing. Aleck was still under great strain and in order to give more time to his experiments he decided to give up his classes. To make matters worse, he was in need of money, and he felt that his clothes were very shabby and down-at-heels. His two friends, Sanders and Hubbard, knowing nothing of his worries over money matters, insisted that he should give up these foolish notions of his about Electric Speech and should get down to work once again on the Multiple Telegraph until he had made a success of it. But Aleck felt that it was much more important that he should clear the ground for his great invention, and he buoyed himself up with the thought that he might at any time stumble on the secret. As it turned out, it was, after all, something of an accident that he did so, and success was much nearer at hand than he dreamed.

CHAPTER TWENTY

The Twang of the Wire

ALECK AND WATSON had two rooms in the top storey of the Charles Williams factory, where they carried on experiments during the day and some times half the night. On a hot day in June, 1875, they were working as usual on the Multiple Telegraph. Watson was in one room operating the transmitter by which the reeds at the receiving end—sixty feet away, in the next room—were made to vibrate; and Aleck was busy tuning up these reeds and listening intently to their sounds.

"The same old thing!" he said to himself, half aloud. "I wish the Multiple Telegraph were out of the way. I'd rather—"

At this moment one of the transmitter reeds ceased to vibrate. Watson, not knowing just what had happened, plucked it to set it in motion, but it seemed to have stuck fast. At the same time Aleck heard a faint "ping" or "twang" come from the wire. He could scarcely believe his ears! That sound could mean only one thing, and with a wild shout he rushed into Watson's room.

"What did you do then?" he demanded. "Don't change anything! Let me see!"

It took Watson only a few seconds to explain what had happened. The make-and-break points of one of the springs had become welded together, and as a result of this interference the electro-magnet was generating a current of electricity—the undulating current that Bell had been listening for, and hoping to find, for nearly a year. He had learned enough about electric currents to know that if this current could make a "ping" or "twang" it could transmit other sounds as well, including the sound of the human voice with all its complex qualities.

Bell was excited. He knew—better than anyone—what the discovery of Electric Speech meant. He knew, though in a dim way, that in that moment there had come into being a new force—and he felt the exaltation that all great discoveries bring to those who share in them.

> "Then felt I like some watcher of the skies
> When a new planet swims into his ken!"

Now, if ever, was the time for the Six Nation's war dance; but Bell was too excited for that, and when Watson had made every move clear to him, the two of them had to go over it all again, and again, and again, and listen to the "ping" a score of times. Never in the history of mankind did the twang of a piece of wire mean so much to the human race! A million of other men might have listened to that faint "ping" without paying any attention to it or without realizing what it meant. But here was a man with a knowledge of anatomy (he knew all about the structure of the ear) and with a knowledge of the science of physics and acoustics, all in one. To him that sound meant one thing. The secret of making wire talk had been revealed.

This electrical machine would never be able to think. It could dream no dreams. It could not take the place of a man's brain. It could not do his will. It was like "the little man" that had tried to speak, in the old house in Edinburgh, but could not. Of the brain of man, a poet once wrote:

> *"This was the home wherein all dreams of earth*
> *And air and ocean, all supreme delights,*
> *Made mirth and madness; wisdom pored alone;*
> *And power dominion held; and splendid hope;*
> *And fancy like the delicate sunrise woke*
> *To burgeoning thought and form and melody."*

But this machine could create none of these things. It could speak no word of its own either to encourage or to dissuade; but when words were poured into its unresponsive ear it could pour them forth again.

The invention of the telephone was not altogether a matter of inspiration, not a flash of lightning out of a cloud, not merely the accident of a passing moment that works in some mysterious way. Bell's instant recognition of the twanging of the wire was the result of long years of patient investigation by study and experiment and clear logical thought. Before this, men had dreamed of speed. "I'll put a girdle round about the earth in forty minutes," said Puck the mischief maker. "In forty minutes!" Was anything ever so utterly preposterous? But today the earth is girdled with sound waves that can travel round the globe itself in less than a single second, less time than a man can take to tell it or even think of it. A miracle has transformed the world!

By means of the telephone—or Electric Speech, whichever you like to call it, mankind can now transcend both time and space, thanks to the twang of a wire in that sultry and dingy attic room!

For the rest of the afternoon and evening the two men, in the stifling heat and with coats off, kept repeating the experiment to make sure that they had not been dreaming when they thought they heard the faint twang of the wire. They tried it over and over with every reed they could find, until they grew weary of the twang. Anyone glancing in through the window would have thought they were playing with a new toy. No one would have guessed that here in this dusty, noisy, hot attic room the world was being remade.

However, this Multiple Telegraph makeshift wasn't a machine for transmitting speech and it was not intended to be put to any such use as this. As yet the "ping" was too faint, and both the sending and the receiving mechanisms must be improved before the tones of the human voice could be carried clearly over the wire.

But now that the undulating current had been generated, it was not so very difficult to make the real talking machine; and before leaving the shop for the night, Bell gave Watson directions for making the first telephone. He set to work at once and his lamp burned far into the night. By the next afternoon two machines were completed, one for the use of each of the men.

The two attic rooms were too close together for a satisfactory test—the sounds from one room could be overheard in the other; and to prevent any interference of this sort, Watson ran a wire down two flights of stairs to his work-bench on the main floor. Then, when darkness fell and the factory was closed for the night, and all was quiet, the two fellow-conspirators came back to try out the machines. Watson connected the telephones with the wires, and then went downstairs to listen, while Bell shouted and sang into the mouthpiece of the new "toy" in the attic work-room. Then Bell put the receiver to his ear to listen for the reply, but heard not a sound, and his heart sank within him. It was a crucial experiment. Would this child of their brain live or die? In that one moment of suspense so much was at stake in this round ball of a world! Then a few seconds later—it seemed an eternity—Watson came scrambling up the steep stairs and burst into the room, breathless and hardly able to speak for excitement.

"I could hear you, Mr. Bell," he cried. "I could almost make out what you said! Almost!"

The two men spent half the night listening to the twanging of the wire and the muffled tones of the voice, and somehow more than once it made Aleck think of "The Little Man," who would go on saying "Ma-Ma" as long as the boys liked to blow the bellows—but who couldn't or wouldn't *talk*.

"To be or not to be!" Bell knew that this machine of his would talk someday, talk too much, perhaps! It is an appalling thing to think of the amount of talking it does, the words that are poured into its ears all over the world from every kind of tongue—such whispering and lisping and laughter, such tears, such low cunning, such madness, such moans, such cries of terror, such passionate pleadings, such whispers of love. When the bell rings it will talk if you talk to it; but yet of all that it says, nothing remains. It is sounding brass and tinkling cymbal! Put all the bell-ringing and all the "hello's" and all the countless nothings over millions and millions of wires, into one swift second, and what a world-maddening shriek it would give forth! "To be or not to be?" The first "talking wire" in the world, in the Williams' attic room was struggling like "The Little Man" to give the answer. It was "to be"; the telephone was in the throes of being born.

That night Aleck sat down and wrote to Hubbard. Hubbard, who had told him to get on with his Multiple Telegraph, and leave this foolish dream! "I have accidentally made a discovery of the greatest importance," so the letter ran, "To-morrow evening I will call to see you and will tell you about it!"

He couldn't telephone Mr. Hubbard to make sure that he would be "in" that evening, and no one knows whether he found his prospective father-in-law at home or what passed between them when Bell was shown into Mr. Hubbard's parlour to tell him of his amazing and exciting and world-shaking discovery.

But whatever may have been said in the course of that interview, Mr. Hubbard was still sceptical, even when this first trial telephone was in actual operation. To his mind, Bell was wasting his time on it which should have been spent on the Multiple Telegraph. As weeks passed into months, Hubbard had become more and more exasperated at the slow progress Bell was making with the telegraph, and at length he told the young man that he must either discontinue his work on the telephone or give up all thought of marrying his daughter, Mabel. It was a desperate position in which to place the youth. He was madly in love—and utterly wretched. Needless to say, Tom Watson had made a shrewd guess as to what the trouble was, although he said nothing about it! Sometimes as they walked, or sat together in the train, there were long intervals when neither of them wished to break the silence; and if at such times Watson had cried, "A penny for your thoughts," he would have found that his hero's thoughts were not concerned with the weather or the train or the Multiple Telegraph, nor with Georgie Sanders!

"I never saw a man so much in love as Bell," said Watson in later life, as he looked back on those anxious troubled years.

CHAPTER TWENTY-ONE

A Crisis

Bell was hard up, but nothing that he possessed could be turned into ready money. He was ready to patent his invention, and it proved in the end to be the most valuable patent in the world. Within two or three short years, had he only known it, he and his friends would count their wealth by millions, all because of that little strand of wire that twanged in the young man's ears in a stifling noisy workshop in Boston on that summer day. But as yet it was just a scrap of paper, nothing more, and it was scarcely worth the parchment on which it was written.

Hard up! No money for new clothes! When he looked at his threadbare coat it gave him a sinking of the heart. How could he appear at the home of the wealthy Hubbards in so shiny and shabby a garb, where he wanted to leave a favourable impression? No money for his landlady from whom he rented rooms in Boston, or for Georgie's grandmama who had opened her house to him in Salem. He had asked the School of Oratory in Boston to pay him ahead of time for lectures that he had not yet delivered, and this money had been used to meet his everyday needs and to discharge his past obligations. He was ashamed to ask Sanders for more loans. Mr. Sanders didn't know, and no one else could guess, what wealth would come to him from his discovery of Electric Speech, so simple a thing that even Bell's best friends would not believe in it. Mr. Hubbard? No, even if things came to the worst he could not go to Mr. Hubbard, whose daughter he wished to marry, and ask for money.

One morning in late August while he was visiting his parents in Brantford, he saw their neighbour, the Honourable George Brown, driving past their door, jogging along into town, down the hill, over the bridge past Brant's Ford, on some errand, no doubt, that had to do with his livestock farm. Three hundred pure bred prize shorthorns! "The man must be wealthy," thought Aleck, "as rich as old Croesus!" And besides this he owned and edited the Toronto *Daily Globe*—"The Scotchman's Bible"—which moulded the opinions of half the people in the Province. The Bells, father and son, went all over the question of money once again, as they had done every day for weeks past, and they finally decided that Aleck should approach Mr. Brown and ask for a small loan. George Brown could be genial, and even jovial, when things were going well; but on the whole he was a rather formidable person, with his lantern-jawed face and his six feet two or more.

Aleck didn't need very much money, two hundred dollars or perhaps three—enough to pay off his petty debts, and a little extra to carry him along and tide him over, to leave him free to put the last finishing touches on his patent,—and then! At the foot of every rainbow lies a pot of gold, but few travellers have learned the secret of how to reach it.

Mr. Brown received the young man cordially, as became a neighbour, and to Aleck's surprise he seemed to be much interested in the new invention; and as soon as the loan was mentioned he offered to advance the money.

"How much do you think you will require?" he asked.

Aleck hesitated. He had thought it all out before he came; but now he was afraid Brown might think he was asking too

much. That amount of money might buy a good thoroughbred shorthorn yearling!

"I should like two or three hundred dollars," he said summoning up all his courage. "Three hundred, if possible."

Brown gazed out of the window. "Three hundred dollars!" he thought to himself "This invention may, after all, be a crackbrained thing. Talking over a coil of wire! There may be something in it, but not three hundred dollars! A talking wire? The thing's absurd! But the young man appears to be intelligent!" Then, turning to Bell, he continued, "Three hundred, for six months, twenty-five a month from me, and twenty-five from my brother Gordon. That makes three hundred. For six months only! I think perhaps we can arrange to let you have the money. But before we can do anything further, I should like to see a copy of the specifications covering your patent. If only the thing were working now it would save me a lot of trouble. This very afternoon I have to go back to Ottawa. Politics and business won't wait, but five minutes over your talking wire would settle it."

He leaned back in his chair and looked at Aleck as if in some hesitation.

"I should like to oblige a neighbour," he continued, "but as a matter of business I can hardly venture on it without some security. No one has any idea whether this machine of yours will stand the practical test. It may not be worth anything. It cannot in any case be offered to the public for some months, or years, perhaps not in my life-time. You have not yet applied for a patent, and there will be delays and not a little risk attached to it. But if my brother and I can arrange to advance you this money, what terms would you suggest?"

Aleck was desperate. He had to have money; otherwise everything would be at a standstill; and a delay in this case would be dangerous. Other men were working on the same problem. If Brown did not advance the money, he must go back to the drudgery of teaching; but, even so, all the time he could spare must be given to the new invention. Every improvement that he might add to it would make his patent, if granted, so much more valuable.

"If you can arrange to let me have the money," said Aleck, "I will make over to you half of the British and foreign rights. But an invention cannot be patented if a patent has already been taken out elsewhere. This means that I have to secure the English patent before I can apply for one in America. I cannot go to England; I have not the money. But you are going over soon. If you will be so good as to visit the British Patent Office on your arrival and have this matter attended to, it will be to our mutual advantage."

Brown agreed to these terms. No formal agreement was drawn up, but Aleck returned to Boston, completed the specifications which he had worked on in Brantford and sent a copy to Brown, fully expecting the loan to reach him promptly, but a fortnight passed, and he heard nothing from Brown. There was nothing for him to do but resume his teaching and deliver the lectures for which he had been paid in advance. October passed, and November, and still no money from Brown.

About this time, a letter from Aleck's mother told of Papa meeting the Governor General at lunch and she added, "The Honourable George Brown was not there." He was amused at a postscript in which she wrote, "By the way as soon as the Sound Telegraph is set a-going, I mean that your father shall send one of his sneezes through it. You would be sure to know who it was!"

Then in November came a great happiness which outweighed his immediate anxieties. In the months of his absence Mrs. Hubbard had helped his cause and now parental objections to his becoming engaged to Mabel were withdrawn. On November 25th he wrote to his mother, "This day has been approved by the Governor of Massachusetts as a Day of Thanksgiving for the Commonwealth and to me it is truly a Day of Thanksgiving, for Mabel has today trusted herself to me and promised to become my wife. This is her birthday. She was eighteen years old this morning. My heart is too full to allow me to write much to you to-night."

By the end of the year Brown had still not been heard from! He had put off his visit to England; and when Aleck went home for the Christmas vacation he sought Brown out at the Queen's hotel in Toronto. He succeeded in renewing the arrangement by which Brown would receive a half interest in all the patents that he would take out for Aleck in England in return for a loan to be paid at the rate of fifty dollars a month—for not more than six months, following the filing of the patents.

Brown did not sail for England until late in January. Aleck saw him off from New York and Brown promised to cable the money immediately on his arrival in England. Three weeks went by, but no cable came. Aleck was heart-sick. He knew, or thought he knew, where the trouble lay. Brown was one of those men—and there were many others—who regarded this "talking wire" as a mere toy. If only it had been shorthorns instead of Electric Speech! Brown put the specifications in the bottom of his trunk and gave no further thought to them. He did not see how they could be of any possible value to Aleck or to anyone else, mere child's play, nothing more. He did not go near the Patent Office, and through his failure to keep his obligations to Bell, he threw away a fortune not only for his own heirs but for the inventor himself.

In the meantime the application for an American patent had to be delayed in order to have the patent taken out in England first. Mr. Hubbard, now at length convinced of the value of Aleck's invention, had become active in promoting it and in safeguarding Bell's interests in it. When nothing was heard from Brown, Hubbard's patience was exhausted. He was Aleck's legal adviser; and without consulting his client, (his future son-in-law), he took the bull by the horns and filed the application in the Patent Office at Washington. And not a day too soon! It was just in the nick of time! Two hours later, Aleck's rival, Elisha Gray, filed what is known as a "caveat," a provisional patent, which would have given Aleck endless trouble if it had come in a few hours earlier. Aleck's application went through the usual channels; and on March 3, 1876, his twenty-ninth birthday, his application for a patent was allowed.

CHAPTER TWENTY-TWO

March 10, 1876

In January, 1876, Aleck gave up his attic rooms in the Williams building. He had begun to entertain suspicions that in the Williams factory with its open workshop his rivals could spy on him and learn something of his projects and plans. After some search he found two suitable rooms in Exeter Place. One of these rooms he used for sleeping quarters, for he had given up his teaching in Salem. The other was fitted up by Watson as a laboratory, and a wire connected the two rooms. They were comfortable quarters, and Aleck was able to carry on his work here without fear of interference or interruption.

During the nine or ten months following the first feeble twang of the talking wire it appeared as if little progress were being made in Electric Speech. The wheels scarcely seemed to move. The new wonder working machine spoke only in muffled tones, with gaspings and splutterings and chokings. Thus far no voice, or sound of any kind, had come clearly enough over the wire. It was only when Bell shouted, in his crisp clear tones, that the faint echoes of words could be heard at the receiving end. All sorts of discs and membranes and wires were used and all sorts of combinations were tried. There was no path for the two adventurous explorers to follow. They were groping, and praying that they might somehow find their way. It was an unknown field where anything, or nothing, might happen. An onlooker might well have begun to think the whole adventure a fiasco. Week followed week, but the two youths kept their courage, coming and going day by day, and making countless adjustments, in the hope that the impossible, or what seemed impossible, might come to pass. Other men had gone through the same heart-breaking experiences over unexplored trails: Galileo with the telescope, Stephenson with the locomotive, Morse with the telegraph, Edison with the arc-lamp, Columbus with the New World.

Then came a certain evening in March—March 10, in the year 1876— a memorable evening for all time. Watson was making ready to try out a new transmitter on which they had been working. It was one in which they were trying to produce a stronger undulating current by using a circuit through a galvanic battery. They had no special reason to suppose that it might be better than a score of others that they had already tried, and even with this new transmitter Bell supposed that he would have to shout aloud as usual to make himself heard; but only through trial and error could they hope to arrive.

At length all was in readiness. Watson was at the receiving telephone and Aleck was at the transmitting telephone in the other room. Watson took up the receiver and was straining to listen for anything that might possibly be different or new, when suddenly there came from the receiver a distinct voice:

"Mr. Watson, come here; I want you!"

Watson could tell from the tone that something was wrong, and he rushed to the other room. Aleck had accidentally spilled some acid from the battery on his clothes and was calling for help. It was the first emergency call by telephone!

But Watson's report on the clearness of the transmission so delighted Aleck that the damage to his clothing was instantly for-

"Mr. Watson, come here; I want you!"

gotten; he rushed to the receiving telephone to listen to Watson's voice. Back and forth between the two rooms they ran, taking turns in speaking and listening. Did they declaim poetry to one another? Probably, for Aleck's mind was well stored with verse, and Watson was accustomed to reciting verses aloud in his efforts to improve his voice. In any case, Watson noted in his diary that first sentence, "Mr. Watson, come here; I want you," as well as some other phrases they used, concluding with Aleck's "God save the Queen." They little knew that within a year or two Aleck himself would be demonstrating his telephone to Her Majesty, Queen Victoria, in person at Osborne!

That night Aleck wrote to his mother, "March 10, 1876—This is a great day with me. I feel that I have at last struck the solution of a great problem and the day is coming when the telegraph wires will be laid on to houses just like water or gas, and friends converse with each other without leaving home."

Today, few people who "get the telephone in" know anything about that exciting evening of March 10, 1876, when the new transmitter proved to be a wonder-worker and the first sentence was clearly heard over the telephone. From that moment forward, the world was no longer the same. It was full of new cries and strange voices girdling the sea and land!

The early telephone consists of a strong ordinary magnet, to the two extremities or poles of which are attached properly insulated telegraph wires. In front of the extremities of the magnet there is a thin plate of iron, and in front of this again there is the mouthpiece of a speaking tube. By this the sounds are collected and concentrated and falling on the metal plate cause it to vibrate. These vibrations in their turn cause electric currents in the two wires which correspond exactly with the vibrations—that is with the original sounds. If now the two wires are connected with an ordinary telegraph line, the sounds can be transmitted to any distance.

This principle is demonstrated in the illustration on this page of a simple transmitter and receiver, and in the illustrations of early telephones on the following page.

(1) the mouthpiece; (2) adjusting screws; (3) a collar and short fixed tube of brass extending to (5) which is the position of the diaphragm, this can be altered by means of the screws (2); over the curled rim of a movable brass collar (4) is stretched the diaphragm (6) which is made of gold beaters' skin; the coil (7) is an electric magnet in which the soft iron rod in the centre projects towards the diaphragm leaving a small space between them; at the centre of the diaphragm on the side removed from the magnet is placed a small piece of clock spring; (8) a pillar holding the magnet in place; (9) a binding screw through which passes a length of telegraph wire (10) from the coil to the receiver; (11) a wire from the battery (12) to the coil; (13) an iron tube inside which is a vertical bar electro-magnet which attracts and causes the thin armature (14) to vibrate.

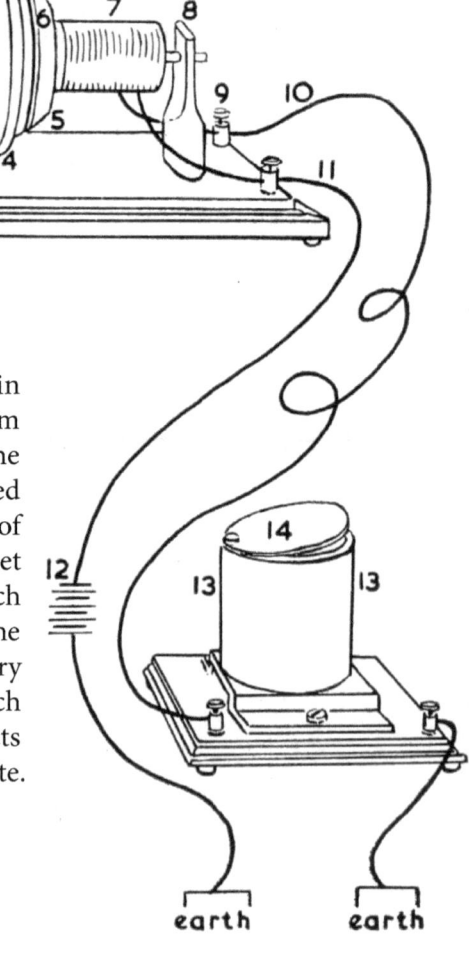

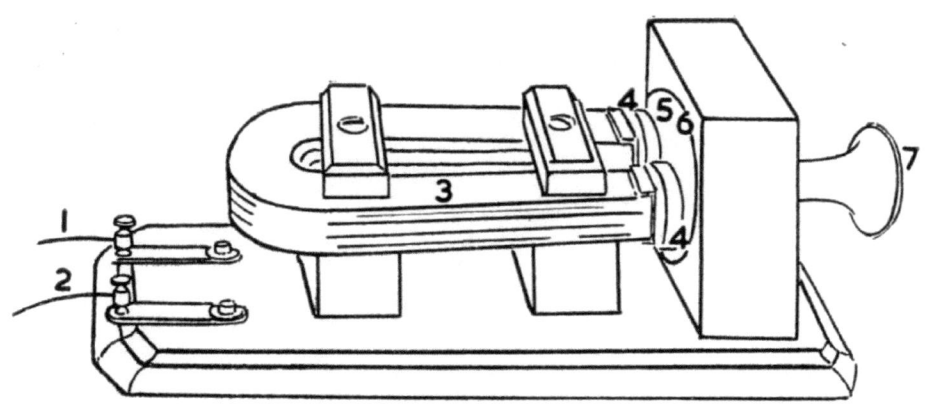

The office telephone.
This model is both receiver and transmitter and the battery is not required: (1) a length of telegraph wire; (2) line to earth; (3) a compound magnet to each pole of which is clamped a short round piece of bar iron, over which is a bobbin of coil wire (4); (5) a small space between the magnet and the diaphragm (6) which is a thin sheet of soft iron; (7) the speaking tube.

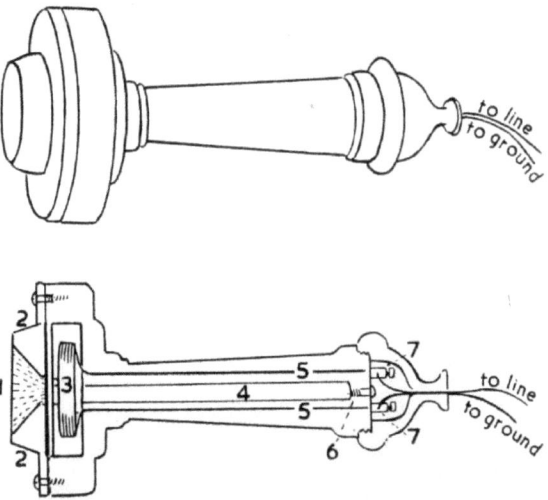

The portable telephone.
(1) a diaphragm of soft iron; (2) the mouthpiece; (3) bobbin of coil round end of magnet (4); (5) wires conducting from coil to binding screws (7). the two wires are then insulated and bound together in one strand for convenience; (6) adjusting screw holding the magnet.

Early Bell telephone 1878.

Edison-Bell phonograph 1886.

A table telephone set of the year 1885.

A table telephone set of the following year 1886.

Table telephone produced by the National Telephone Co. about 1895.

The Stranger automatic telephone.

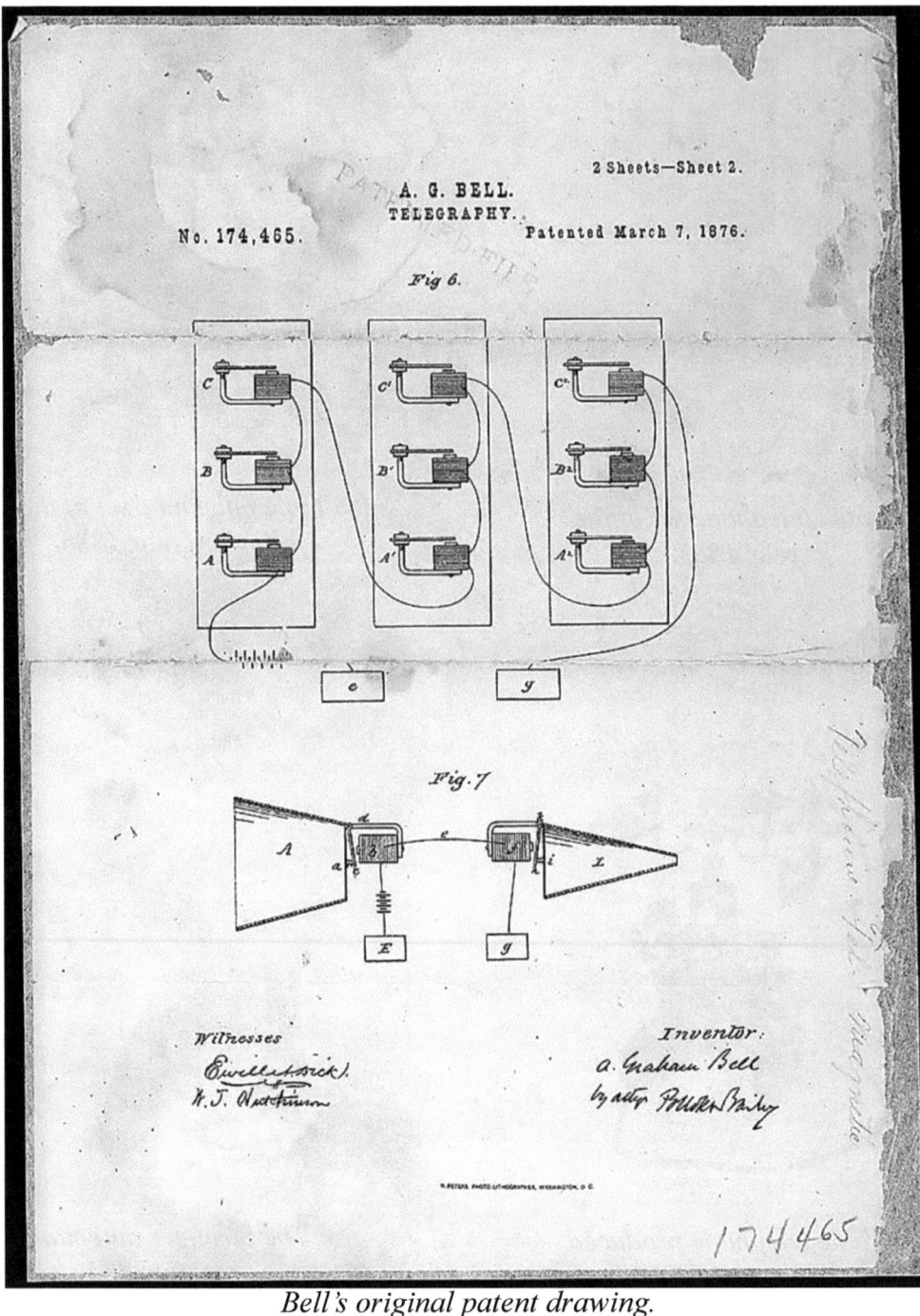
Bell's original patent drawing.

Part 4.

BRIDGING SPACE.

AT LAST! AT LAST!
A Perfectly Reliable Acoustic Telephone.

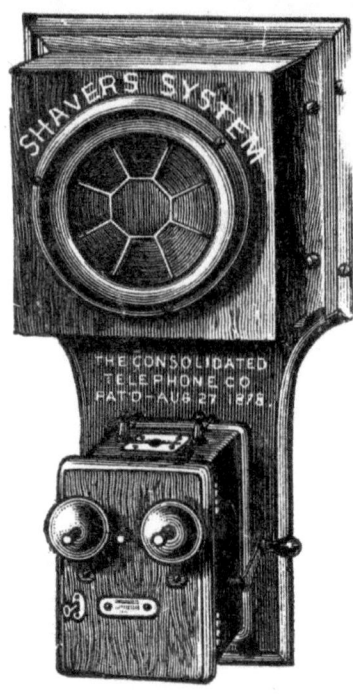

A TELEPHONE THAT WILL WORK WELL OVER ANY ROUTE, REGARDLESS OF ANGLES.

A TELEPHONE THAT WILL WORK WELL IN ALL KINDS OF WEATHER, WIND OR RAIN.

A TELEPHONE THAT DOES NOT ROAR WHEN THE WIND BLOWS.

A TELEPHONE THAT WILL ADMIT OF SEVERAL INSTRUMENTS UPON THE SAME TRUNK LINE.

An Acoustic Telephone Central Office,

Adapted to switch one line with another, similar in effect to the electric system. This is attained by the SHAVER SYSTEM, and its practical utility demonstrated by its use by over 150 bona fide subscribers in New York City, who are paying rentals of $2 to $8 per month for the service.

TELEPHONES sold outright or rented in unoccupied territory. Send stamp for list of users and testimonials. It will pay you to investigate the truth of these statements, and secure an agency. Liberal discounts to agents, and satisfaction guaranteed or money returned. References.

To rapidly introduce our goods, we will, for the first line in any town, give a discount of sixty per cent. from our regular retail prices for Telephones, as given below, providing the purchaser will endeavor to secure us a reliable agent.

Retail prices Telephones each	$10.00
Magneto Call Bells (not necessary)	6.00
Wire per 100 feet	.25
Hangers for right angles	.25
Ordinary supports	.10

The SHAVER SYSTEM of Telephony has been in use by many business houses in this city for the past two years, gives universal satisfaction, and we cheerfully recommend it to those desiring such service. —FRANK LESLIE'S PUBLISHING HOUSE.

THE CONSOLIDATED TELEPHONE CO.,
Jersey City, N. J.

An ad for an early telephone.

CHAPTER TWENTY-THREE

The Emperor

It was one hundred years since the Declaration of Independence had been signed; and the people of the United States planned to celebrate the occasion with a great exhibition—the centennial—to be held in Philadelphia in that hundredth year—the summer of 1876. So much can happen in one hundred years! There was, for instance, the invention of the electric light, the sewing machine, the telegraph, the reaper, and a thousand-and-one other labour-saving devices. And only a few months before, a teacher of the deaf, a young man named Bell, had invented an instrument for talking to people at a distance over a wire!

At first Aleck did not think he would take the trouble to prepare to exhibit this new "toy"—an "American humbug"—the London *Times* had called it. He felt that he needed more time to make improvements on it; and by the time he had made up his mind to enter it among the electrical exhibits, where it properly be longed, it was too late, and it would not have been entered at all if Mr. Hubbard had not found a place for it among the school exhibits of Massachusetts. The result of these delays was that it was squeezed into a narrow space under a stairway. There it remained for the next six weeks, without attracting the attention of anyone. Late in June Mr. Hubbard telegraphed to say that the judges were expected to reach the electrical exhibit on Sunday,

June 25; and Aleck had to decide whether he would go to the Exhibition or not. Mabel Hubbard had decided not to go, and she had difficulty in persuading Aleck that he should be there to look after his exhibit. He knew that the telephone would not attract attention, and would receive no award, unless he was there to demonstrate it; but he felt it his duty to remain in Boston to look after the examinations at the School for the Deaf, at the close of the term; and it was not until the last moment, when the train was ready to pull out that she was able to prevail on him to go. But it seemed as if the fates had conspired to prevent the newly-born telephone from having a "fair show." The weather suddenly turned so hot that it left everyone, visitors to the Fair, judges, and the inventor himself, limp and jaded; and what everyone wanted most was to go home to escape the torturing heat.

It was late on the Sunday afternoon before the tired judges reached the electrical section, with the school exhibits just beyond. The general public were not admitted on Sundays; but a large group of privileged guests had been attached to the judges' party, and progress was slow.

The group of judges included Sir William Thomson, (Lord Kelvin), the most eminent scholar of his time; Joseph Henry, whose encouragement had meant so much to the young inventor; Elisha Gray, Bell's keenest rival, and numbers of others.

Among the guests, the most distinguished was Dom Pedro, Emperor of Brazil, with the Empress and their retinue. Dom Pedro was the guest of the nation, and, as such, he was treated with great deference. He was tireless, and went everywhere, bent on seeing and hearing everything. The afternoon wore on. It was growing late. The judges were nearing the end of their labours for the day, and word was given out that the next exhibit, the one that immediately preceded Bell's, was to be the last. The guests breathed a sigh of relief; but Aleck, who had observed what was taking place, from his alcove above, felt a sinking of the heart. This decision meant that he would not have an opportunity of demonstrating his telephone and could have no hope of award.

But when the judges had completed their work and the company were about to disperse, something dramatic and unforgettable happened. Dom Pedro caught sight of a tall, dark-eyed, pale young man who was standing by the stairs in the section beyond. There could be no mistake! And with a glad cry of recognition he rushed over to him with outstretched hands. The Emperor had visited the School for the Deaf in Boston some weeks before and had talked at length with Mr. Bell, the instructor, about the education of deaf children. A charming young man whom the Emperor could not soon forget!

"Why, how do you do, Mr. Bell," the Emperor exclaimed. "I'm so glad to see you again. How are all your deaf people getting along, up in Boston?"

"Very well, I hope, Your Excellency," Bell replied. "I have run away from them today, but I must get back to Boston to-night. Their examinations begin tomorrow. They can't say anything about this weather, whatever they may think of it."

"I wish you were staying longer," said the Emperor. "I should like to have more talk with you."

"I cannot possibly remain," said Aleck, "but I should like to ask a very great favour of Your Excellency. I have with me here an electrical talking machine, which I should like you to see, if you can spare a few minutes before you go. I shall not have another chance to demonstrate how it works."

Ah," said Dom Pedro, "then we must have a look at it now."

To the judges and guests who waited, out of courtesy, till Dom Pedro should leave the building, this interruption was puzzling.

Who could this young man in a threadbare coat be, who was on such familiar terms with the Emperor? That was a question they could not answer, but they must, perforce, see the play through to the end.

The telephone was in position, and Aleck hurried off to take his place at the transmitter, five hundred feet away. Then when all was ready, he began to recite "To be or not to be," which his grandfather had taught him during that first year in London. "To be or not to be!" His grandfather's instruction served him in good stead on that day!

While Aleck was reciting, each of the listeners at the receiving end, one after the other, took turns in pressing the iron box of the receiver to his ear, and each of them repeated the phrase which he himself had heard. Then Sir William Thomson, Chairman of the judging committee, said, "I will go and talk," and a moment later the next listener, Elisha Gray, heard Sir William's voice repeating, "Ay, there's the rub." He in turn repeated it to the group standing by who broke into applause, and with that he passed the receiver over to Dom Pedro, who listened with an expression of doubt, which gave way to utter amazement as he repeated the now familiar words, "To be, or not to be?"

Among the judges the new telephone created something of a sensation. They forgot all about the heat; and neither guests nor judges were in a hurry to hasten away.

"It is the most marvellous thing I have seen in America." said Sir William Thomson—high praise from the foremost scientist of the time. And when Joseph Henry listened and heard the tones of the human voice coming over the wire, an expression of something like awe appeared on his face.

When the demonstration was concluded, the judges and guests, Dom Pedro included, gathered round the young man to shake hands and congratulate him on his marvellous achievement; and numbers of others stayed to listen, and to talk over the wire, until the buildings were closed for the night. The next morning the telephone was removed to the judge's pavilion so that they might hear it again and demonstrate it to their friends; and from time to time during the day, Sir William Thomson and Lady Thomson ran to and from the transmitter to the receiver like two delighted children.

In the meantime Alexander Graham Bell had taken the train back to Boston, to return to the humdrum tasks of the schoolroom on the following day. This was the proudest day of his life.

CHAPTER TWENTY-FOUR

Stovepipe Wire

WHEN THE JUDGES had made their award, the Centennial was a thing of the past, as far as Bell was concerned. A fortnight later he returned to Brantford, to enjoy a well-earned rest, and to share his triumph with his father and mother at Tutelo Heights. They in turn were proud to share it with the friends they had made among the townfolk. The "queer" young man had come into his own. They were interested in young Mr. Bell, the inventor, and they knew now why he was making faces at himself in a mirror and shouting "Oh" and "Ah" all day long.

But there were still a good many of the good folk of Brantford who didn't know what it was all about. It was a strange enough thing, this idea of talking to people a long way off, by means of an iron box and coil of wire! "And heaven knows," said one old lady, "there's plenty of folks at a distance I'd just as soon not talk to at all." This "contraption," some folk thought, couldn't amount to much anyway, for it was so small a thing that it was all carried in the travelling bag which the young man had brought with him.

Professor Bell, of course, knew all about this wonderful invention and as soon as Aleck reached home, father and son began to plan how they should try it out. The verandah, they agreed, was the place for the transmitter, and the small stable close to the edge of the river-bank was far enough away for the receiver.

The coil of wire that Aleck had brought with him connected the two, and it provided the resistance that was required. The thing did talk, but not very clearly, and they had to shout into it to be heard. It was like the little voice of a lost child—but it talked, and that was all that really mattered.

Then they set it up in the house and ran the wire from room to room in what Aleck called "parlour experiments"; and when they tired of that, Aleck ran a wire around the house under the eaves and sat in his room sending messages to himself. The next evening some of the neighbours on the Heights were asked in. This time the wire was run out to a grapevine arbour in the grounds and the neighbours talked into the machine and listened to their hearts' content. A soloist from one of the city churches, who lived near by, was asked to sing into the telephone, but could not think of anything appropriate, except her last Sunday's solo, "I Need Thee Every Hour," which, everyone agreed, was quite in keeping with the occasion. The neighbours, of course, could not call one another up in the morning and talk things over; but the time for that sort of thing would come soon enough!

These experiments thus far were all carried on with transmitter and receiver on the home grounds. But it was still a question whether messages could be heard over a wire for a longer distance; and the Bells, father and son, decided to try it out between Brantford and Mount Pleasant, a little village two or three miles farther along the main road. For this purpose it was necessary for them to have the use of the Dominion Telegraph Company's lines between the two places. This was granted and it was arranged that Aleck should listen at the receiver in the Mount Pleasant office while his Uncle David Bell recited Shakespeare and sang songs into the transmitter at Brantford. At the appointed time Aleck waited nervously in the telegraph office, with watch in hand. At first there was blank silence, then a preliminary cough, and then a voice repeating the well-known lines, "To be or not to be; that is the question!"

The question, as far as the telephone was concerned, was already answered. It was to be!

A few days later, Professor Bell decided to hold a public reception at his home, "Melville House," to which the townspeople would be invited, to do honour to the inventor and give him an opportunity to demonstrate his invention; and the arrangements were again made with the Dominion Telegraph Company in Brantford that the Bells should have the use of the telegraph wire from Mount Pleasant to the city for one hour. But it was a quarter of a mile from the Bell home out to the main road where they could connect with the telegraph wire; and without a wire to bridge this gap, there could be no way of sending messages to the Bell home from Brantford. But if the worst came to the worst, Professor Bell planned to run a short line from the verandah to the stable or to the grapevine arbour, as they had done before.

But Aleck proved equal to the occasion. He drove out to Mount Pleasant, taking the transmitter with him, to make sure that it was possible to talk over the telegraph line to Brantford; and then he drove into the city and bought up all the stove-pipe wire in town. On his return home he asked his neighbours, Mr. McIntyre and Mr. Brooks, to help him to string the wire along the rail fence from the Bell home to the main road. Mr. Brooks was drawing in grain, and he suggested that they wait until the afternoon so that he might finish his work in the field.

Then the three men set to work with tacks and staples; but they had no means of insulating the wire. At one point they came to a culvert under which the wire had to be run; and a small boy (a lad named Leslie) who was looking on, crawled in and pushed the

wire through. It was something for a boy to have helped to build the first long-distance line in the world! And it was something for him to remember in after days, that the inventor himself brushed the dust and mire from his clothes!

It did not take very long to string the wire and to Aleck's surprise and delight, although the stove-pipe wire was not insulated it gave no trouble at all. Mr. Brooks and Mr. McIntyre must have wondered what this performance was all about. Some of the other neighbours nearby were curious as to what was going on, and they watched Aleck and his helpers from their windows.

"I've heard tell of many strange things," remarked one old lady, "but anything to beat a man stringing a wire through the country to talk through it is the silliest piece of tomfoolery ever was. He's clean daft."

But the demonstration, and the singing and reciting, went off well. For the singers Aleck had provided a telephone with a triple mouth-piece, and three young girls sang a three-part song into the transmitter to demonstrate it to the guests. The idea that more than one person could speak, or sing, over the same wire at one time, was more wonderful than anything else they had yet heard.

At the reception there were a hundred guests or more, and the narrow country road was blocked with horses and carriages of the townsfolk, and it seemed more like a garden party than a demonstration of this new and unknown miracle that was to change men's ways of living and bring the ends of the earth within their reach.

But the real test of long-distance talking came a few days later. For a try-out it seemed to Aleck and his father that nothing could be better than the line from Brantford to Paris, eight miles away, where their friends the Hendersons lived. For this experiment the Dominion Telegraph Company once again gave the use of their lines. It was arranged that Uncle David Bell, who was now living in Brantford, should speak through the transmitter while Aleck was at the receiving end in Paris; and Aleck drove to Paris that afternoon, taking the iron box receiver with him. It was a pleasant drive over the quiet country road, past farms on which the harvesting was now in full swing, where wagons piled high with sheaves were busy "drawing in," and the fence corners were gay with aster and golden-rod, and the swamps were lined with Joe Pye weed. It was a lovely countryside, and the rolling uplands covered with pines added to the picturesqueness of the scene.

At Paris, Aleck went to the telegraph office, connected his telephone wires with the Company's line, then took up the receiver and waited. But when the appointed moment came, all he could hear was a crackling and whistling—"static," we call it now—and voices of speakers and singers so faint as to be scarcely audible. It was not possible as yet to hold a two-way conversation over the telephone. It was a one-way machine. So Aleck telegraphed to Brantford, suggesting a change in the coils, and with that the voices of singers and speakers came through quite clearly. But Aleck was puzzled by one thing. He thought he heard the voice of his father, who was supposed to be absent on another engagement. For his own satisfaction, he sent another telegram of inquiry, and was told in reply that the speaker was his father, who had been able to take part in the conversation after all.

This day, August 10, 1876, was a red-letter day in the history of the telephone. This was the first long distance message ever given or received over a real wire. It was an event not only in the life of Alexander Graham Bell, but in this planet of ours. From this time forth all the earth was to be bound together in a network of wires. But when Aleck Bell drove home that evening through the quiet countryside he thought only of the events of this one day. Even for him it must have been hard to picture the million miles of poles and wires and the thirty million homes in which the telephone has proved its worth, the telephone, the servant of mankind, carrying sounds from a far distance to the ends of the earth.

"What do you read, my lord?" asked old Polonius.

"Words, words, words," quoth Hamlet; and from this day to the end of time, it was and will be, words, words, words—and still more words!

CHAPTER TWENTY-FIVE

The Shuttle

There was no doubt now that the telephone could talk. The trouble was to get it to talk back. It was not possible then to hold a "conversation" over the wire. You could, if you wished, use a double set of transmitters and receivers and carry on a conversation by turning from one to the other, or you could do as Aleck did at Paris when he first talked over "long distance"—send your messages and questions one way by telephone, and wait for the answer to come back by telegraph, or vice-versa. But that was very inconvenient; and as soon as Bell returned to Boston, he and his friend Watson set to work to try to solve the problem of a two-way conversation, a sort of shuttle, over a single wire.

As usual, the solution of the problem proved to be very simple, when once it was discovered! It was found that if thin metal discs were attached to the membrane of each of the diaphragms so as to form patches, the same wire could be used for either talking or listening. The experiments which Bell and his friend Watson carried on were first tried out in their Exeter Place rooms, where they were able to work without interruption; but when at length after two months of experiment they were able to carry on a two-way conversation to their own satisfaction, they decided to try out their new device over a real line. Among their friends were the Walworth Brothers, manufacturers, with offices in Boston and factory at Cambridgeport, two miles away. Bell readily obtained permission to try out his experiment over the Company's private telegraph line which connected office and factory. They decided, of course, to wait until after dark to make the test, when the factory should be closed, and all quiet, and when there should be no danger of discovery. It was agreed that Bell should wait at the Boston office for a signal from Watson, who would talk from the Cambridge end of the wire; and that evening at the appointed time Watson sallied forth with the telephone under his arm and carrying a coil of wire.

At the factory he was admitted grudgingly by the night-watchman, who looked upon him as a suspicious character who was worth watching. When the telephone was connected up with the telegraph wire, Watson put the receiver to his ear and shouted into the transmitter and waited for Bell's reply. "But," to quote from Watson's own story, "not a sound could I hear, although I knew that Bell must be shouting! I shouted into my instrument and listened again. Nothing but black dismal silence! I looked over the connections, readjusted the telephone, then listened and called again. Silence still! The thing was absolutely dumb!" Something had gone wrong. The Walworth electric current was sufficiently strong. No doubt of that. Could it be that there was a leakage of electric power, which weakened the current? That could be ruled out, but something was wrong. No sound came from the wire. He shouted again, and again. No reply! He was sick at heart. There would be nothing else for it but to return to Boston, a weary two miles, and carry the unwelcome news of failure to Bell.

He tried the line again and checked all the connections. It was like shouting into the dead man's ear! He took the telephone down, wrapped it up in his newspaper and prepared to set out for the city. Then suddenly, in a flash, another possibility occurred to him. Was there another wire in the factory, a relay which drew off the power? The only way to find out that was to trace the wire from the point where it entered the factory. He did so, scanning the wires by the light of the watchman's lantern. In the end the search led him to a locked office. The night-watchman opened it unwillingly, and there on the wall before him was the switch that was causing the trouble. He shut off the current and went back to his telephone. His first call, the first ever made over a two-way telephone, was answered. More clearly and distinctly than he had ever heard it between the two rooms came Aleck's voice, "Where have you been all this time?"

In order to have proof of their success in talking intelligibly over a wire two miles long Watson and Aleck had arranged to record their conversation. Each was to write down what he said and what he heard and the two records, put side by side, would convince the sceptics. This they did, but when they had written down what they thought was enough for their purpose, they continued their conversation in sheer delight that they could do so when two miles apart. Watson induced the doubting watchman to listen, too. And then he wrapped up his precious telephone, wire, and tools in a newspaper and hurried back to Exeter Place.

There the two celebrated in an Indian War-Dance duet, for Watson had learned that accomplishment from Aleck. In their glee, they forgot there were other people in the house. Next morning, the landlady stopped Watson on his way out and said, "I don't know what you fellows are doing up in that attic but if you don't stop making so much noise at night and keeping my lodgers awake, you'll have to quit them rooms." Said Watson, in recounting the incident, "She wasn't at all scientific in her tastes and we were not prompt with our rent."

CHAPTER TWENTY-SIX

In England

IN THE NEXT few months Bell and Watson made further improvements in the telephone and carried out other long distance experiments over telegraph wires. But the two men, Mr. Hubbard and Mr. Sandders, who were paying the cost of these experiments, began to wish for some financial returns; and Bell, too, was anxious for money to enable him to get married. And so they made an offer of all Bell's patents to the Western Union Telegraph Company for one hundred thousand dollars. The company promptly, and some what scornfully, refused. A few years later it would willingly have paid twenty-five million dollars for the patents; but they were no longer for sale!

Meantime, as the public learned more about it, the telephone created quite a "stir" among thinking people everywhere, and Bell received invitation after invitation to explain and demonstrate his invention in public. In Salem he described the telephone to an eager audience of five hundred people, and to prove to them how it actually worked he called up Watson at his workshop in Boston, eighteen miles away, and asked him to talk over the wire to his Salem audience. Then someone in Boston played a cornet solo, Watson sang "Hold the Fort," and Bell himself sang "Auld Lang Syne" into the transmitter. An account of the lecture was telephoned from Salem hall to a reporter in Boston and the next morning the Boston Globe published the first newspaper despatch ever received over the telephone. The unbelievers rubbed their eyes and read the despatch again. Henceforward the telephone was something that had to be taken into account.

During that winter, Bell lectured in several towns and cities, including New York. Distinguished scientists, the poets Longfellow and Holmes, and President Eliot of Harvard, were among those who listened to him. His courteous manner and fine presence on the platform, his trained, resonant voice, and his power of clear explanation made him an outstanding speaker; and his use of musical instruments and songs as well as speech to demonstrate the telephone provided an entertaining evening. The money he received for these lectures helped somewhat to solve his financial problems. It was the first money earned by the telephone.

In the spring, an agent was ready to undertake the establishment of the telephone in England. On the strength of these prospects Bell won Mr. Hubbard's consent to his marriage, and in July, the wedding took place. Immediately following the event the bride and groom took a train to the old homestead on Tutelo Heights to spend a few days with Aleck's father and mother before setting out for England.

The young wife had heard a great deal about Brantford and for her, as for her husband, it was a joyful home-coming. They walked around the little farm, stood on the bank looking down over the "rapid river," sat beneath the birches in Aleck's "dreaming place," and walked out along the country road, where the stove-pipe wire was still hanging to the fence rails.

Within a few weeks they sailed for England. The honeymoon was spent in Scotland in a little cottage near Elgin, the scene of young Aleck Bell's first experiments with vocal sounds. Then they began housekeeping in Kensington, London. Mrs. Bell, young and beautiful, devoted her keen intellect and her great talents to helping her brilliant husband. She wrote his letters, taking his dictation by reading his lips—and in every way was a real helpmeet.

And, indeed, Bell needed her help, for he was plunged that autumn and winter into a busy round. Various learned societies honoured him by calling meetings for the special purpose of hearing him demonstrate his telephone. Great scientists such as Professor Tyndall and Sir William Thomson were among those present in these assemblies, and there was not a vacant seat. He lectured, too, on the education of deaf children, and in addition to all this, he was experimenting on methods of overcoming the interference which his telephone lines suffered when close to other lines. And, of course, he had to give attention to the formation of the Electric Telephone Company for which he superintended tests which were to attract English capital to the venture.

In January came a red-letter day when he received a "command" from Queen Victoria to demonstrate the working of his telephone at Osborne House, on the Isle of Wight, where Her Majesty was then staying. This visit had to be talked over beforehand. He had to learn how to meet, and talk to, Her Majesty; and he had to make his description of the telephone so simple that the Queen, who was not familiar with scientific terms, could readily understand it. Like a good teacher he made the construction of the machine clear to her by the use of models; and among other devices he showed her the two halves of a telephone that had been cut down

the middle to show how it was constructed. Then, of course, the Queen herself had to talk through the telephone, and she held a conversation with the Princess Beatrice and with Sir Thomas Biddulph, Master of the Queen's Household, at Osborne Cottage near by, and listened to "Kathleen Mavourneen" and "Coming Thro' the Rye," which were sung over the telephone. That evening in her diary the Queen spoke of having seen and heard the telephone. "A Professor Bell," she added, "explained the whole process, which is the most extraordinary."

But on the heels of all this excitement—lectures, demonstrations, experiments—there followed those inevitable vexations that every new invention of this kind brings in its train. As soon as it became evident to the rabble of fortune-hunters—disappointed inventors and others—that there was a fortune in this new "toy," they began to lay claim to the invention, most of them without a shadow of evidence to support their pretensions. Legal firms fought their battles for them in the hope of sharing the spoils if they should be successful; and new companies were formed who disputed Bell's claims. The Electric Telephone Company was faring badly and Bell was disgusted with the whole business of trying to establish the telephone as a paying concern in England.

But in May, a new happiness crowned his domestic life. A little daughter was born to the Bells. The delighted Papa wrote to his mother, "She popped into the world with as lusty a shout as ever demonstrated the possession of a healthy pair of lungs. Such a funny little thing it is! Perfectly formed, with a full crop of dark hair and bluish eyes and a complexion so swarthy that Mabel says she has given birth to a red Indian! I can't say much about good looks. I never could see beauty in a baby, but she is our own baby and that is enough for us."

Before long, cables from home informed Bell that the powerful Western Union Telegraph Company was manufacturing telephones which infringed his patents and that the Bell company were about to sue the company in the courts. Cable after cable came, asking him for papers required as evidence, and by autumn an urgent demand was made that he return to Boston to help with the suit. He was heartily sick of it all and he replied that he was through with the telephone and was returning to teaching. He would leave England in October, he declared, but not for Boston. He would take his wife and child to Brantford, instead!

It was Watson who won him round. Watson met the boat at Quebec, accompanied the Bells, baby daughter and nursemaid, to Brantford, and persuaded Bell to leave his family there and return with him to Boston. Bell prepared his testimony but the trial was not completed, as the company, seeing they could not win, made a settlement.

When Bell found that the telephone business was now more promising, he gave up his idea of returning to teaching for a livelihood. At the request of the Bell company he remained with them, for a time, on salary, for this lawsuit proved to be only the first of many. In the years to follow, more than six hundred suits came before the courts, involving Bell's rights to his invention. The specifications for his patent, over which he had toiled so painstakingly in order that, as he said, there should be "no possible hole" in it, stood firm against all the legal attacks massed against it, and every one of the suits was decided in his favour.

CHAPTER TWENTY-SEVEN

The Telephone, Past and Present

THE DEVICE THAT made it possible to carry on a "two way" conversation over the telephone was only one of a multitude of improvements that were made during the life-time of the inventor. Since that evening in Exeter Place when Watson heard Bell's cry of distress, "Mr. Watson come here. I want you," many changes and adjustments have been made, each one contributing in some way towards making the instrument more efficient. "I cannot claim what you know as the modern telephone," said Bell a few years before his death. "It is the product of many, many minds."

In 1876, those who talked or sang over the new telephone, had to shout to make themselves heard; and two years later when Queen Victoria listened to this new "toy," she mentioned in her diary that the voices sounded very faint and that she had to press the receiver close to her ear in order to made out what was being said. But now when you talk from London to Glasgow, or New York to San Francisco, or from Halifax to Vancouver (which are 4,200 miles apart), you can hear as clearly as if the speakers were in the same room.

"Mr. Bell," said Issawa, a Japanese student, at one of those early lectures on the telephone in the centennial year, "Can this thing talk Japanese?" And for answer Bell told him to bring along another Japanese and try it for himself. Issawa found out that it does not matter what sounds you make or in what language you speak on the telephone, so long as there is someone to listen to what you say.

In 1876 there was not any bell or buzzer to call you to the telephone, and at first the only signal that the listener could hear was a faint knocking sound when the speaker tapped the machine with his pencil to attract attention. Watson's first improvement on the lead pencil was a signal which became known as 'Watson's Thumper." His next device produced a harsh screech and was called "Watson's Buzzer." He finally hit on the idea of using a polarized bell. This was so successful that the principle on which it worked is still used in present day bells.

In those first days of the telephone, when you wished to greet the person who called you to the telephone, you shouted "Ahoy! Ahoy!" as if you were about to embark on a voyage; and your friend at the other end of the line called back, "Ahoy! Ahoy! Are you there?" But there came a time when someone, no one knows who, started calling, "Hello!" and by and by at a telephone convention, the two parties, the "Hoy! Hoy!" fans and the "Hello!" fans, took sides and put it to the vote to decide which call-word should be used; and feeling ran so high that the one party had badges printed bearing the battle-cry "Hello!"

In 1876 there was no such thing as a telephone booth; and when Watson tried to cover up the instrument so that other people would not overhear, he used a blanket as a sort of tent, under which he sweltered and smothered while the talk went on. But now the speaker can talk from his home or his office or in a booth, or even out of doors, or from a moving train, without any inconvenience or discomfort.

Nearly forty years after the first conversation between Bell and Watson over the two-mile wire between Boston and Cambridge, they held another eventful telephone conversation. But this time it was across a distance of four thousand miles, from New York to San Francisco. It was at the opening of the first transcontinental telephone line, in January 1875. They heard one another as clearly as they had done over that first two miles of space. When the ceremonies of the afternoon were over, including a conversation between President Wilson in Washington and the Governor of California in San Francisco, Dr. Bell spoke again to Watson. He asked Watson to wait a moment until he should attach another transmitter which he wished to try out. In a few moments Dr. Bell's voice came as clearly and loudly as before.

"I am now talking through a duplicate of the first telephone you made for me and that we tested together in June, 1875. Do you hear me?"

Watson replied that he heard him perfectly. Then came from Dr. Bell, "Mr. Watson, come here. I want you."

"I should be very glad to, Dr. Bell, but we are now so far apart it would take me a week to come instead of a minute," Watson replied.

"He heard me," said Bell, "but he did not come immediately. It's not long now, however, before men will be able to appear from across the continent within a few hours after they are summoned."

CHAPTER TWENTY-EIGHT

At Home and Abroad

By 1880, Bell had ceased to take an active interest in the telephone. His mind was full of other inventions. As Mrs. Bell wrote, more than forty years later, "He never did talk much about things after they were done. His interest always was in what he was doing at the moment. That always was the most important, even when nobody else thought so."

The family moved to Washington and there he began experiments in producing and reproducing sounds by means of light. "I have heard a ray of the sun laugh and sing! I have been able to hear a shadow," he wrote.

In the late autumn of 1880, he went to Paris to receive the Volta prize of fifty thousand francs, awarded to him for the invention of the telephone. On his return, early in 1881, he put the prize money to work, by establishing with it a laboratory which he called the Volta Laboratory. There, with two associates, he continued experimental work. One of their achievements was the invention of the first successful gramophone record.

In the next year, 1882, Dr. Bell became a fully naturalized citizen of the United States. His citizenship was a matter of great satisfaction to him, and in later years he used to say to his grand-children, "I am a better American than you are; I chose American citizenship but you were only born that way."

In the meantime the Bell family circle in Washington was increasing. A second daughter was born and was described by Bell in a letter to his mother as "a real, substantial fact, a kicking, screaming, undeniable fact. At birth she roared so loud that she was heard at Mr. Hubbard's house in K Street, a full block away, by telephone. But now she has become reconciled to the idea of becoming a peaceable and much admired member of society—and rarely breaks the silence with more than a grunt, which nurse translates in the word, 'supper,' a verb, active, in the imperative mood."

For some time Bell had been trying to persuade his parents to come to Washington to live and in the early eighties they did so. "Our only comfort and stay now is in you," his mother had written many, many years before, after the loss of her other sons. And a "comfort and stay" to his mother he proved to be to the end of her days. Writing to her when he was almost fifty years of age he said, "I wish indeed I could make it a happy New Year for you, for I do love you just as much as when I was a little boy in your lap, and my heart is with you now, though I am a greyheaded man."

Washington now held all that was dearest to Bell, his wife and daughters, his parents and his work; and for the remainder of his life it was his home. He was one of its most distinguished citizens—a Regent of the Smithsonian Institution and one of the founders of the National Geographic Society.

Although Dr. Bell had too many compelling interests to permit of his spending time in trivial social affairs he became famous as a host. There were no social events in Washington more interesting or more distinguished than his "Wednesday Evenings." At these weekly gatherings he cordially welcomed to his home scientists, teachers, and authors. These occasions were not merely formal receptions. In the handsome library of his home, Bell, by his vigorous, stimulating personality and his kindliness and courtesy created an atmosphere in which such men were at their best. He put his guests at their ease and drew them out to talk of their interests. And always in the midst of the gathering Dr. Bell kept an honoured place for his father, Professor Melville Bell, whose training had first aroused in the son the interest in speech which had resulted in his great invention of the telephone.

And in these years Dr. Bell was by no means forgetting the cause of the deaf. Now that he had more money and more time at his disposal, he began elaborate research into the question of heredity in deafness. He renewed his efforts to improve the education of the deaf. Bell felt strongly that every deaf child should be taught to speak instead of to use signs and should associate with those who could hear instead of being isolated with only the deaf for companions. In 1888, his work for the deaf brought him recognition in Britain. He was to go there and speak before a Commission appointed to inquire into methods of educating the deaf. Internationally known as an inventor and scientist he always preferred the title, "a teacher of the deaf."

CHAPTER TWENTY-NINE

Beinn Bhreagh—Beautiful Mountain

PROFESSOR MELVILLE BELL had for a long time wished to revisit Newfoundland, where he had spent several years in his younger days; and as it happened, his son Aleck, in moments of leisure, had been reading a book written by that charming New England author Charles Dudley Warner, about the Bras d'Or Lakes in Cape Breton, Canada—Bras d'Or, "Golden Inlet." Gold did not mean much to Bell—except the gold of sunset. He had sufficient money now to indulge himself if he wished, and he wanted to visit Cape Breton and the Bras d'Or country and follow in the footsteps of Charles Dudley Warner. Mr. Hubbard too wished to visit Cape Breton to look into the coal mines at Glace Bay, in which he had money invested.

And so it came about that in the year 1885, two or three years after their return from Britain, the Bells set sail for Newfoundland, intending to visit Cape Breton and the little town of Baddeck on their voyage northward. They reached Baddeck safely, and spent a few days at the village inn, "The Telegraph House," from which Warner had explored the countryside and the lake. Then they set out for Newfoundland; but before they had gone far, their small craft was shipwrecked; and they returned to Baddeck to make a longer stay. The village and surrounding country and the Golden Inlet charmed them.

It seemed just the place for the kind of living they desired for their little daughters, away from the restraints of fashionable summer resorts. Two sons, Edward and Robert, had been born but both died as infants and the two little girls were, if possible, more precious than ever to their parents. Mrs. Bell was far in advance of her time, for she wished her little girls to run about freely, unhampered by petticoats and full skirted dresses, dressed as boys, preferably. Here, in Cape Breton, they could live exactly as they pleased, with the benefit of salt water and cool summers. The Bells decided to return the next summer.

Across from Baddeck Bay, to the south and east, there rises a bold promontory, nearly one thousand feet in height; and in their second summer here, Bell and his wife rowed across the bay to the point, and from there explored the hillsides and the clearing on the brow of the hill. From this open space they looked out over a glorious panorama of lake and wood and distant hills with farmsteads covering the lower slopes of the mountain. Here they determined to establish a permanent summer home.

They planned a cottage, enthusiastically made a cardboard model of it, and had it built on the first land they were able to acquire, half-way up the slope. But from year to year, as opportunity came, Bell bought up the neighbouring farms, until at length he came to own the whole beautiful Point. He then built a spacious and beautiful permanent summer home on the tip of the promontory so that from its windows one looks out, as if from the prow of a boat, over the surrounding waters of the Bras d'Or lakes. They named their home "Beinn Bhreagh," Gaelic for "Beautiful Mountain," and no name could have better described it.

Year after year honours from the outside world came to Alexander Graham Bell. He received the French Order of Merit, the Legion of Honour; honorary degrees were conferred by Universities; learned societies elected him to membership; a school with special classes for the deaf was given his name. And he travelled abroad from time to time, once giving a year to a trip around the world. But henceforth neither honours nor travel meant as much to him as his summer home, "Beinn Bhreagh." It was an ideal retreat from the passions and strife of the outside world. And it was in this island of Cape Breton with its rockbound, surf beaten headlands and salt arm of the sea gleaming in molten gold, that he found another "dreaming place."

CHAPTER THIRTY

"No End to Striving"

FROM THE TIME when the Bells first let down their anchor in Baddeck Bay until the end of his life, was a period of more than thirty years. The full story of Bell's activities during these years would fill a large volume. He had now enough money. Someone has said that the three most notable documents in the world are Magna Carta, the American Declaration of Independence, and the Bell patent on his telephone. But the value of this patent is not to be measured in dollars and cents alone. It too was a charter that helped to set the human spirit free; and the wealth that it brought was not a treasure to hoard but to spend freely for the benefit of mankind.

A laboratory was, of course, a necessity to Dr. Bell and one was built at Beinn Bhreagh soon after he acquired the property. There he worked, day in and day out, far into the night, on an amazing variety of projects. His investigations and experiments had to do with these subjects among others: waste of heat from open fireplaces, distilling salt water, the breeding of sheep, hereditary deafness, mines and rocket guns, submarines, and a hundred other things. "Don't keep forever on the public road," he once said, "going only where others have gone and following one after the other like a flock of sheep. Leave the beaten track occasionally and dive into the woods. Every time you do so you will be certain to find something that you have never seen before;" and Bell himself practised what he preached.

First and foremost he became intensely interested in the problem of aerial flight. In his younger days he had startled Watson by his assertion that someday men would fly. He was now convinced that the day was near at hand when the problem of aviation would be solved.

"We know perfectly well," he said in an address delivered in the year 1914, "that the time is coming, is almost here, when it will be an everyday thing to go from place to place through the air. A man proposes to try this summer to fly across the ocean in a heavier-than-air flying machine. The strange thing about the matter is that the experts who have examined the possibilities, find that he really has a fighting chance."

For a time Bell thought that he might solve the problems of mechanical flight by making use of the principle of the schoolboy's kite; and for over ten years he experimented at Baddeck with kites of various sizes and shapes, made up in red silk—"queer thing-a-ma-jigs" a local boatman called them. He finally evolved a tetrahedral kite, that is, one with four triangular faces. Some of these were twelve to fourteen feet in size and had an upward pull strong enough to pull a man thirty feet up into the air.

In 1907, Dr. Bell had with him at Beinn Bhreagh four young associates—Glen Curtiss, a builder of motor-cycles, "Casey" Baldwin who had just graduated from the University of Toronto as a mechanical engineer, Mr. J. A. McCurdy, still a student in the same university, and a trained observer, Lieutenant Selfridge of the United States Army. All of these young men were enthusiastic believers in the future of aviation and each was an expert in his own line.

As always, Mrs. Bell was following her husband's work with

her keen mind; and, at her suggestion, and with the assistance of twenty thousand dollars which she placed at their disposal, this group was formed into the Aerial Experiment Association. Bell was completely happy—a master with this gifted group more or less his "pupils."

The story of all the attempts these experimenters made to "get into the air" would require a volume in itself. They met with disappointments at first. But with the third machine they built, Curtiss won a trophy by flying it successfully on the first measured official flight in the United States, July 4, 1908. With their next machine, McCurdy, at Baddeck, in 1909, made the first successful flight in the British Empire. And Baldwin and Bell, using some of the discoveries they had made in their experiments with flying machines, together perfected the "hydrodrome," a high speed boat, which in 1910 was the fastest boat yet designed. And it was this group which first used pontoons on aeroplanes for starting from and alighting on water—the forerunner of modern seaplanes. And Dr. Bell's ardent support of aviation, in public addresses, as well as in every other possible way, at a time when people in general were scornfully sceptical, was in itself a force for the advance of aviation.

Meanwhile, life at Beinn Bhreagh was not all scientific work—not all aviation. Mabel Bell had from the beginning of her life at Baddeck taken an interest in the women of the neighbourhood. In 1891 her husband, writing to his mother in Washington said, "Mabel is the most indefatigable woman I have ever seen. She is a born organizer." And he went on to tell of the eighty girls she had organized into sewing classes, the products of which were sold in Montreal, and of a Literary Society she was planning to form.

And the family life at Beinn Bhreagh was not allowed to suffer from the demands of either science community service. Dr. Bell did not remain aloof from the fun of the life of a summer place. They explored Cape Breton, using horse-drawn vehicles and a houseboat,—"Mabel of Beinn Bhreagh," in their early days on the island. This houseboat, in after years, was beached and became an ideal retreat for Dr. Bell. There he would go on a Saturday and remain until Monday, when he wished to be alone to think out some scientific theory. No one ever disturbed him there.

As the daughters reached their teens their father enthusiastically directed them and their friends in amateur theatricals. He loved a ceremony, a festivity, and birthdays and similar events were celebrated joyously. In 1894 his parents celebrated their golden wedding at Beinn Bhreagh, his mother now a frail little lady of eighty-five with whom her grand-daughters were warned not to be "rough." She had lived to see her one remaining son, for whose chance of health she had ventured all a quarter of a century before, a hale and hearty man, covered with honours. Three years later she died in Washington.

The passing of time brought sons-in-law to Beinn Bhreagh, Dr. Gilbert Grosvenor and Dr. David Fairchild, each to become distinguished, the former as editor of the National Geographic Magazine, and the latter as a plant explorer. Ten grand-children rounded out the happy family circle. And what happy memories those grandchildren have of Dr. Alexander Graham Bell as "Grampie!" He must not be disturbed in his laboratory, but at five he would join the family for tea and fun. He told the children wonderful story serials, continued from summer to summer. One of these followed the fortunes of the man who was made of rubber, who could blow himself up and who was once carried away to the South Seas by a hurricane! Another "serial," had to do with the giant, "Imagination," who could lengthen or shorten his legs at will, according to what he wanted to do with them!

At the dinner table "Grampie" had other stories. He would have in his pocket a clipping from something he had read, or a memorandum of something he had saved to tell the family. With his animated face and eloquent gestures the stories lost nothing in the telling. But he did not like to have anyone leave the table to speak over the telephone! He had not invented the telephone that people should be so bad-mannered as to do by means of it, what they would not dream of doing in person—disturb people by a call during the dinner hour!

Then there were the jolly evenings around the piano when he played the accompaniment to rollicking choruses in which everyone must join. And, true to his unfailing consideration of his wife's deafness, he would have her place her hand on the piano so that through feeling the vibrations she might share in the music. Theirs was a long and happy love story! In his endeavours her enthusiastic trust and belief in him was his greatest inspiration.

Alexander Graham Bell making the first telephone call between New York and Chicago in the year 1892.

Alexander Graham Bell with his wife, Mabel Gardiner Hubbard, and their daughters Elsie (left) and Marian.

CHAPTER THIRTY-ONE

The Reminder

At the intersection of four streets—King, West, Albion, and Wellington—in the City of Brantford, Ontario, "The Telephone City," there stands a monument to Alexander Graham Bell—a "reminder" of the inventor and his work, for that is what the word "monument" means.

But if this monument were nothing more than a mere "reminder" it would not need to be here at all; for in every office, and in almost every home in the land, there is a telephone, an every-day reminder on the desk or on the wall, that rings when someone wishes to call you, whether from a neighbour's house or from some "station" half-way round the world. Listen to its persistent clanging! "Mr. Watson, come here. I want you." That is what it said on the first day when it spoke to the world! It was a humdrum message, a cry of distress, but, after all, the telephone is a sort of servant, that ministers to our "wants."

On the hilltop to the south and west of the city of Brantford, where the busy town has crept out to meet the suburb of Tutelo Heights, stands the modest farm house which was once the Bell home ("Melville House," the Bells called it), with neat lawn and flowers and broad driveway leading up to the door. It, too, is a "reminder," for it was here that the idea of the telephone first took definite form. It was here that it was "conceived," if not

born. The City of Brantford has set this homestead aside as a perpetual memorial!

But neither the old homestead on the heights, nor the telephone itself, with its millions of poles and endless miles of wire, is a "monument" in the ordinary sense of the word. When you look at a service-box with its transmitter and receiver you think of the telephone merely as a machine. Not one of these instruments—not even the one that was presented to His Majesty the King—is a thing of beauty. They are intended only to serve you when you are "wanted on the phone." But the Bell monument on the by-street of this ambitious little town, is intended not only to remind you of the invention but to awaken within you a feeling of exaltation, a sense of triumph at the march of mankind towards its goal:

"The Soul that from a monstrous past,
From age to age, from hour to hour,
Feels upward to some height at last
Of unimagined grace and power."

When you have crossed the little park towards the monument, walk up the broad steps that lead to the central bronze panel, and study it. That reclining figure at one end of the panel is Man, and bending over him is the figure of Inspiration urging him to greater effort; and at the other end, floating in the air are Knowledge, Joy, and Sorrow, which the telephone has brought to mankind. And those heroic female figures on the great pedestals at either side represent humanity sending and receiving a message.

The monument had been eight or nine years in the making, and now at last it was complete and ready for the unveiling. The day that had been fixed for the ceremony was October 24, in the year 1917, at the hour of high noon. But the weather man was out of sorts on that morning and sent showers instead of sunshine. In spite of this, the city folk were in holiday mood, and rain could not dampen their spirits. No school that day! Crowds lined the streets. The band of the Duffer in Rifles played lively airs. Soldiers in khaki mingled with the crowds, for the First World War was then at its height. Notables were there from near and far, Members of Parliament, the Mayor, the Lieutenant Governor, Chief Hill of the Six Nations, and officials who were "high up" in the telephone world. The Duke of Devonshire, the Governor General of Canada, was there to address the assemblage and unveil the memorial. Alexander Graham Bell, his name a household word in every corner of the earth, was there to tell the listening world anew the story of his invention, Alexander Bell of Edinburgh and Elgin and Bath and London, and of Brantford and Boston and Washington and Baddeck too,—and of all the world. Alexander the Great! And Mr. McIntyre and Mr. Brooks, honest farm neighbours, who had helped to string the first long-distance wire, had left their ploughing for the day, to join the crowd. And last, but not least, Walter Allward, the sculptor who designed this striking "reminder," was there, not to speak, but to let his monument speak for him!

The ceremony at the Square did not take long, for there was no protection from the weather, and Jupiter Pluvius would have his way! In spite of the rain a crowd had gathered in the little park in front of the monument, waiting patiently for that brief moment when the draperies would fall from the great bronze figures at either end and reveal the whole design in all its fine proportions.

The Governor General performed that part of the ceremony, and then the crowd made their way to the Opera House where they might be able to listen to the preceedings in comfort.

The Governor General spoke briefly about the "miracle" of the telephone and the part it had come to play in everyday life.

"It is," he continued, "indeed a memorable day, not only for Brantford but for humanity, and the ceremony in which we have taken part will live for many, many generations after we have all passed away, and future generations will be proud of the part we have taken.

"I have already formally unveiled the monument. I now formally dedicate it and hand it over to the City in trust for all time to come."

The inventor himself was the next speaker. When he came forward to speak, the vast crowd rose to their feet, and such shouts and cheers, such rounds of deafening applause the grey old walls had never re-echoed before.

"Your Excellency," he began, "Ladies and Gentlemen. There are some things worth living for and this is one of them. I came to Brantford in 1870 to die; I was given six months lease of life, but I am glad to be alive today to witness the unveiling of this beautiful memorial that has been erected in the City of Brantford. As I look back upon it, visions come to me of the Grand River and of Tutelo Heights and my dreaming place upon the heights, where visions of the telephone came to my mind. I little thought in those days that I should ever see a memorial like this—a memorial that is not only gratifying to me personally as an appreciation of my own personal effort to benefit the world, but is an appreciation of the invention itself.

"Much of the experimental work of the development of the apparatus was done in Boston. Still I am glad to be able to come forward and say that the telephone was invented here."

This, indeed, was the point that the inventor made clear, that here, on Tutelo Heights, the idea of the "talking wire" first took definite form. Here many of his early experiments had been carried on; and here the first draft of the specifications covering the patent had been drawn up by the inventor himself. But it does not really matter, after all, where the telephone was "invented." Brantford, in Canada, and Boston, in the United States, both contributed something. In Canada, the inventor found a place where he could rest quietly and think his idea, his "dream" of electric speech through to the end; and in the United States he found the opportunity to experiment, to put his theories to the practical test. There is no boundary line between Old Ontario and the United States of America as far as the telephone is concerned. It does not belong to either; it belongs to mankind.

Later in the afternoon Dr. and Mrs. Bell and members of their family drove to Tutelo Heights and spent "a sad and happy hour" in wandering about the grounds and visiting the familiar rooms in the old home; and among the landmarks of interest were the birch trees under which he had rested in his hammock—his dreaming place, in which the problem of the undulating current and the "talking wire" came to him in its earliest form.

His own mind was busy that day, with thoughts of the people and events of the past—his illness, his parents who had hazarded their all for him, his music, his labour on the farm, and that first perilous drive to town, his Paris friends, the Mohawk chiefs (children all!), the dead man's ear (or woman's, it mattered not), that he had assailed with shouts and cries; the curious, inquisitive, well-meaning folk, who thought him a little "queer;" and George Brown with his six feet two and his thoroughbreds, and his Scottish "burr." And—how could he forget it?—the old river below the high bank with its birches and his "dreaming place." It could tell a story too of the days when life was leisurely, and there was no telephone on the wall to wake one with its insistent ringing.

Three years after the unveiling of the monument at Brantford, another honour was conferred on Alexander Graham Bell which touched his heart more than any other that had come to him. He was given the freedom of his native city, Edinburgh. The boys of his old school, The Royal High School, loudly acclaimed its famous Old Boy that day. And the councillors of Edinburgh, seventy-two in number, clad in scarlet robes, stood while the Lord Provost made the presentation, addressing Bell as "Alexander Graham Bell, Esquire, Doctor of Philosophy, Doctor of Laws, Doctor of Science, Doctor of Medicine."

Alexander Graham Bell

CHAPTER THIRTY-TWO

At Close of Day

At the time of the unveiling of the monument in Brantford, Alexander Graham Bell was seventy years of age—a very old man he seemed to all the young people who got a holiday from school that day. And to Bell himself it seemed a long, long time to look back to that day sixty years before when he turned the paddle wheel to brush the husks off the wheat at Herdman's Mill, so much had happened in those world shaking years. But he was still young in heart, and life was still something to be enjoyed.

> *"How dull it is to pause, to make an end,*
> *To rust unburnished, not to shine in use,*
> *As tho' to breathe were life!"*

"What a glorious thing it is to be young," he once said to a group of young people, "and have a future before you. But it is also a glorious thing to be old and look back upon the progress of the world during one's own life-time."

And there were still a few years left! Five years more, and then he too heard far voices calling him, and word went abroad to the ends of the earth that the Master was dead. It had been a busy life and a useful one, but it was time to rest. He was buried from his home at Beinn Bhreagh, "the Beautiful Mountain," and the funeral rites were simple, as he himself wished. The coffin was covered with twigs of spruce from the hillside, woven together by his grand-children. It was borne from the house to the hilltop, a mile away, in a buck-board drawn by his beloved bays. Members of his own staff were the pall-bearers. The service reflected the character and tastes of the man. There was no sermon, no fulsome eulogy, nothing that was formal or insincere, just The Lord's Prayer, two simple hymns, the 91st Psalm, Bell's own commentary of Longfellow's PSALM OF LIFE; and the first verse of Stevenson's "Requiem."

> *"Under the wide and starry sky,*
> *Dig the grave and let me lie.*
> *Glad did I live and gladly die,*
>   *And I laid me down with a will."*

He was buried at sunset in the cleared space of "Beautiful Mountain," on the high point looking out over the Bras d'Or Lakes—the golden arm of the sea.

> *"Here—here's his place, where meteors shoot, clouds form,*
> *Lightnings are loosened,*
> *Stars come and go! Let joy break with the storm!*
> *Peace let the dew send!"*

A great boulder of granite marks his resting place fronting the sea.

At half-past six in all the stations from sea to sea, which bore the sign of the bell, "the talking wire" fell silent for the space of sixty seconds, while the little group standing with heads bared beside the open grave spoke no word; and as the light of sunset fell across Beinn Bhreagh, for one brief moment the babel of voices from near and far was stilled.

Printed by Libri Plureos GmbH in Hamburg, Germany